MANAGING CANAL IRRIGATION IN INDIA

About the Author

Sib Ranjan Misra, a Ph. D. in Economics (Jadavpur University) and D.E.P. (Istituto di Studi per lo Sviluppo Economico, University of Naples), was on the teaching staff of the University of North Bengal (1971-78). He is currently engaged as a Professor of Economics in the Department of Economics and Politics, Visva-Bharati University, Santiniketan. His present assignment includes research and teaching in various branches of Economics. He has completed quite a number of projects on problems of tea plantations, fisheries development, irrigation management etc. Sponsored by ICSSR, UGC, ICAR etc. He has also served for quite some time as an Advisor on the Fisheries Development Advisory Board, Ministry of Agriculture and Co-operation, Government of India. The author's publications have appeared in various professional journals and also include books entitled '*Tea Industry in India* (1986), *Fisheries in India* (1989), *Floriculture in West Bengal – Problems and Potentials* (co-authored, 1997).

MANAGING CANAL IRRIGATION IN INDIA

Problems and their Resolutions

SIB RANJAN MISRA
Department of Economics and Politics,
Visva Bharati University, Santiniketan

CONCEPT PUBLISHING COMPANY, NEW DELHI-110059

ISBN 81-7022-814-X

First Published 2000

Published and Printed by
Ashok Kumar Mittal
Concept Publishing Company
A/15-16, Commercial Block, Mohan Garden
New Delhi- 110059 (India)

Phones: 5648039, 5649024
Fax: 091-(11) - 5648053
Email : publishing@conceptpub.com

To my mother,
now a cherished memory
who first introduced me
to many wonderful things
in my life

'Sangachhdwam Samuadhwam
Sahaveeryan, Karava-Vahai'

—*The Rig Veda*

This means :

'Move together, talk together
Together we will rise and flourish.'

PREFACE

The growing need for increased agricultural production, growing scarcity of cultivable land, highly erratic and inadequate rainfall and availability of modern inputs encourage irrigation development on a rapid scale. India has been investing on irrigation development at a rate unmatched by any other country in the world. It is also true that India has an enviable record of engineering accomplishments in the field of irrigation and these accomplishments rightly deserve much of the credit for the country's impressive growth in foodgrain production since its Independence. But unfortunately, the problems like increased construction cost, getting favourable locations for projects and poor utilization of the country's already-created potential have shifted the emphasis from construction to management. Canal irrigation offers comparatively higher scope for such management than well and tank irrigation. This is because each canal benefits a large section of the community, offers high management intensity and there is vast unutilized created potential. It is also possible to increase both equity and productivity in canal irrigation system. It is held that the performance of major canal irrigation system is poor where the objectives and goals are not realized.

After about fifty years of experience with canal irrigation system in the post-Independence period, there is a view emerging where it is held that canal irrigation has proved a costly experiment. We are, in fact, very much in the dark as to the nature and distribution of the effects the canal irrigation has upon the economy in general. Farmer's participation in irrigation management is probably as old as irrigated agriculture itself. But knowledge about farmer-managed irrigation system, traditional and new, is scattered and the subject has not been adequately researched. Till recently, many canal irrigation projects have been considered, designed and managed almost exclusively from an engineering point of view. The socio-economic, environmental and other dimensions do not seem to have been integrated with the technical designs. The researchers are still at large to identify the reasons for the emergence of less efficient irrigation system. Hence, how to ensure equitable and efficient allocation of available water among crops or users has been a major concern for the planners in the recent past. In the absence of well-articulated rules to ensure efficiency in irrigation water distribution, maldistribution tends to take place with all its attendant socio-economic and environmental consequences. This study makes an attempt, in a modest way, to examine the factors responsible for the unsatisfactory performance of canal irrigation projects in our country.

This book which runs into seven chapters has a structural layout as follows. The need, scope and potential for development of irrigation in India are introduced in Chapter One, where also the objectives and the methodology of the study are spelt out. In Chapter Two, attempt has been made to critically evaluate the basic problems that this sector in general has been facing. Chapter Three sets out with an introduction of the Mayurakshi canal irrigation system, the study area and presents micro measures of the impact of canal irrigation on the farm economy of the command area. The functional analysis which precisely highlights the contribution of canal irrigation in terms of regression co-efficients, production elasticities etc. has also been presented. 'Area effect', 'crop pattern effect', 'cropping intensity effect', income distribution effect' etc. resulting from provision of canal irrigation facilities have also been examined. Chapter Four gives an account of the factors leading to conjunctive use of water and the problems and the constraints experienced by the farmers. Chapter Five gives an account of the status of water distribution and examines some of the issues associated with water management in irrigation projects. This study clearly brings out the fact that only technical and administrative measures cannot provide the remedy. It shows that an understanding of the socio-economic aspects of irrigation is necessary. This book in its Chapter Six addresses itself to the problems of people's participation in irrigation management. Issues like arguments in favour of people's participation, what it is, why people are reluctant, why people should co-operate/participate, how it affects the entire operational system have been discussed. It shows that participation of the beneficiaries in irrigation management is imperative. Chapter Seven contains the conclusions and policy implications thrown up by the study.

Of course, micro-studies, like this, by nature cannot be blown up for any overall general estimation. But the plus point of such problem-oriented micro-studies is that one has an opportunity to have a close view of the problem.

I owe a special debt to all those irrigation officials and farmers at the village level, mostly very poor and fatigued, who cheerfully bore with me in responding to all of my umpteen and tedious questions. These have provided useful insights into the basic problem. I sincerely thank them for their constructive comments. I also place on record the help and co-operation I have received from Dr. Madan Gopal Ghosh, the ex-Director of the Agro-Economic Research Centre and the service colleagues at the Department of Economics and Politics, and Agro-Economic Research

Centre, Visva-Bharati University. I warmly appreciate and acknowledge the help rendered by Dr. Debashis Sarkar and Shri Nihar Mukhopadhyay whose troubleshooting abilities with the computer have been particularly helpful in organising and printing the manuscript. Last but not the least, I would like to express my heartfelt thanks to my family members for providing support and emotional sustenance over the years.

Department of Economics and Politics
Visva-Bharati, University
Santiniketan, W.B.

SIB RANJAN MISRA

CONTENTS

LIST OF TABLES

ONE

NEED, SCOPE AND POTENTIAL FOR DEVELOPMENT OF IRRIGATION IN INDIA

In India, the importance of irrigation in accelerating economic growth in general and agricultural development in particular has been widely recognized and accordingly, a high priority has been assigned to it in the successive development Plans. The average annual outlay on major and medium irrigation projects including CAD (Command Area Development) has increased from around Rs. 760 million in the First Plan to nearly Rs. 56,000 million in the Annual Plan 1995-96.[1] Similarly, for minor irrigation, the average annual outlay has increased from Rs. 130 million in the First Plan to Rs.15,200 million in 1995-96. During this period (1951-95), total public sector investment of Rs. 580,000 million was made for major, medium and minor irrigation development including the CAD Programme in India. As a result of such massive investment, the area under irrigation increased from 22.60 million hectares, at the beginning of the First Five Year Plan (1951) to about 77.77 million hectares at the end of March, 1995. The share of groundwater development is about 52 per cent in the total irrigation utilization and about 50 per cent in the total irrigation potential created upto the end of March, 1995. Foodgrain production in India has increased from about 55 million tonnes in 1949-50 to about 191 million tonnes in 1994-95.

But certain important considerations emerged, India is likely to be subjected to water stress. India's population is estimated to be 1,086 million in the year 2007 and 1,394 million in the year 2025 and accordingly, per caput water availability per year will come down to 1,920 m^3 and 1,496 m^3 respectively (Navalawala, 1997).

Per caput foodgrain crop land decreased from 0.18 hectare in 1981 to 0.15 hectare in the year 1991 against an average of 0.19 hectare of South Asian countries. Out of its total geographical area of 329 million hectares, the total area available for cultivation is about 142 million hectares without accounting for fallow lands and 165.65 million hectare including fallow lands. The average size of operational land holdings has reduced from 1.69 hectare in 1985-86 to 1.57 hectare in 1990-91. Although the combined percentage of marginal (below 1 hectare) and small (1-2 hectares) land holdings is 76.2 per cent of total number of land holdings, its percentage to total operated area, nevertheless, comes to just 29 per cent. As per 1981-82 Land Utilization Statistics and Reports till 1984-85, the availability of net cultivated land per caput decreased from 0.48 hectare in 1951 to 0.20 hectare in 1980 and will further decrease to 0.15 hectare by the turn of this century. Since 1970-71, the net sown area has remained stagnant at around 140 million hectare, but cropping intensity increased from 118.2 per cent in 1970-71 to 130.16 per cent in 1992-93, mainly as a result of extension of irrigation facilities. Out of the total net sown area, about 127 million hectares have been under cultivation for foodgrain crops for the last two decades. Further, there will be pressure on good cultivated lands for uses other than agriculture due to urbanization, industrialization etc. As such, it is likely that land available for cultivation of foodgrain crops could come down to about 122 and 120 million hectares by the year 2007 and 2025 respectively and in that case, per capita foodgrain crop land would further reduce to 0.112 hectare and 0.086 hectare respectively.

Again, it should be mentioned that in spite of impressive success in raising foodgrain production, the annual compound growth rate of agricultural performance has remained moderate vis-a-vis population growth. Overall agricultural productivity (in 1993) in the case of foodgrain crops in India is about 2 tonnes/hectare as compared to other countries like China, South Korea and Indonesia which have a productivity of 4.58, 5.60 and 3.91 tonnes/hectare respectively and a global average of 2.74 tonnes/hectare.

Another important consideration in the strategy for water resource development in India has been the need for production of food, fodder, fibres, fuel and urbanization. All these put severe competing claims on landuse as well as on water use as a result of population growth. Again, so far water use for domestic purposes has been low as a result of the low level of development and low per capita income. Higher level of development results in higher income level which, in turn, increases the demand for domestic water. Today, on an average, it is about 80 litres per caput per day for the total population. This is due to the low level of access to safe drinking water and sanitation. Although, in India, 80 per cent of the total population (1991 Census) have access to safe drinking water, the population that have access to sanitation is a mere 14 per cent.[2] Surely, the water demand per caput of total population will have to be increased to at least 120 litres per day by 2007. Further, it will be required to consider the water demand for livestock population which could be about 3 to 4 km^3 annually. Taking all these factors into account, broad estimates for water demand in the years 2007 and 2025 for the projected population and food self-sufficiency would show that the water demand will be of the order of 967 km^3 and 1,362 km^3 in the years 2007 and 2025 respectively. Again, due to competing demands for water from other sectors, particularly domestic and industries, the share of irrigation from the total water demand will have to be brought down from its present level of 84 per cent to 74 per cent in the year 2025. This will only be possible when the water use efficiency (which is at present 40 per cent) in irrigation is improved to the level of at least 60 per cent. Unless the water use efficiency improves to a reasonable extent, the lack of fresh water is likely to be a major physical constraint for future food production and development in India.

The frequent occurrence of droughts and famines in the country has heightened the need to harvest available water fully; particularly for, rainfed farming is fraught with multiple risks. Farmers operating in arid and semi-arid areas under rainfed conditions face, more than most others, a multitude of perceived and oft-experienced risks of varying severity emanating from the uncertainties inherent in their natural, economic and socio-economic environments. These risks mainly arise from the natural environment and uncertainties in climate. It is in this context that irrigation becomes important to provide insurance against possible risks in farming, particularly rainfed. Several forms of irrigation interventions for drought resistance have widely been heralded.

However, India has an enviable record of engineering accomplishments in the field of irrigation and those accomplishments deserve much of the credit for the nation's impressive growth in foodgrain production since its Independence.

When India started planned economic development in 1951, the irrigation potential from major and medium irrigation sources was about 10 million hectares and from minor irrigation works was 13 million hectares, totalling 23 million hectares (Table 1.1). At the end of 1995–96, the total irrigation potential from major and medium works increased to 33 million hectares and minor works contributed 56 million hectares making a total of 89 million hectares. The ultimate irrigation potential from all irrigation works was estimated at 114 million hectares a few years back[3] (Table 1.1).

As far as irrigation programmes are concerned, RIDF-I (Rural Infrastructure Development Fund-I) with a corpus of Rs. 2,000 crores was launched in 1995–96 to attend to the needs of rural infrastructure which aims at accelerating completion of ongoing irrigation projects.

Table 1.1 : Outlays on Irrigation and Irrigation Potential Generated in the Plans 1951–97

Plans	*Outlay/Expenditure (Rs. crores)*	*Irrigation Potential Cumulative (Million Hectares)*	*Irrigation Utilisation (Million Hectares)*
Pre-First Plan	——	23	23
First Six Plans	16,030	68	61
Seventh Plan	17,530	77	69
Eighth Plan	32,530	90	80

Source : Eighth Five Year Plan (1992–97), Vol. II, Chapter 3, Economic Survey, 1996–97.

During 1996–97 also, an amount of Rs. 2,500 crore was provided under RIDF-II. In 1996–97, a programme called "Accelerated Irrigation Benefit Scheme" (AIBS) was launched under which the Centre is providing additional assistance by way of loans to the States on matching basis for early completion of selected large irrigation and multipurpose projects. Rs. 900 crore was provided to the States under this scheme in 1996–97. The 'Approach Paper to the Ninth Plan' lays emphasis on improving the efficiency of end-use of water through adoption of water efficient devices, promoting conjunctive use of surface and ground water, ensuring progressive reduction of conveyance losses in irrigation schemes, strengthening the organisation of Command Area Development etc. (GOI, 1996). Firstly, as a consequence of irrigation, less than 18 per cent of net cropped area was irrigated area in 1950–51. This has increased to 34.5 per cent in 1991–92 (Table 1.2). Further, there has been a gradual improvement in area irrigated more than once. Whereas area irrigated more than once was only 8.1 per cent of net irrigated area, it improved to 22 per cent in 1970–71 and was about 31 per cent in 1991–92.

Table 1.2 : Gross and net Irrigated Area in India

(Million Hectares)

Year	*Net Area Irrigated*	*Gross Irrigated Area*	*Area Irrigated more than once*	*Net Irrigated area as per cent of sown area*
I	*II*	*III*	*IV*	*V*
1950-51	20.9	22.6	1.7	17.6
1970-71	31.1	38.2	7.1	17.6
1991-92	48.8	64.0	15.2	34.5

Source : Agricultural Statistics at a Glance (1995), pp. 79–80.

In absolute terms, area irrigated more than once has gone up from 1.7 million hectares in 1950–51 to 15.2 million hectares in 1991–92. Area irrigated more than once is a kind of land augmentation.

Secondly, canal irrigated area has increased from 8.3 million hectares to 16.9 million hectares during the last four decades. Even then, its relative importance has come down from 40 per cent to 35 per cent. During this period, it is the well irrigation which has made the most spectacular progress. In 1960–61, only 0.1 million hectares were irrigated by tube-wells; but, in 1990–91, over 14 million hectares by tube-well irrigation. Tube-wells account for 30 per cent of total irrigated area. Well irrigated area has increased from 6 million hectares to 24 million hectares during the last 40 years. Well irrigation in 1990–91 accounts for 51 per cent of the total irrigated area as compared to only 29 per cent in 1950–51. The growth of well irrigation has been at the expense of irrigation from tanks and other sources.

Table 1.3 : Area Irrigated by Sources in India

(Million Hectares)

Sources of Irrigation	*1950-51*		*1990-91*	
	Area	*Percentage*	*Area*	*Percentage*
Canals	8.3	40	16.9	35
Wells including Tube-Wells	6.0	29	24.1	51
Tanks	3.6	17	3.2	7
Other Sources	3.0	14	3.2	7
Total	20.9	100	47.4	100

Source : CMIE, Basic Statistics Relating to the Indian Economy (1994), Table 6.4.

Thirdly, it is also evident from Table 1.4 that the total net irrigated area increased slightly more rapidly (2.4 per cent per year) during the period 1968-83, essentially after the Green Revolution than during the first period 1951–68 - before the Green Revolution. Among different types of irrigation, growth of canal irrigation command area proceeded at an even pace - about 2.3 per cent per year - during both periods. Tank area grew at a similar rate of 2.4 per cent during the first period and then declined at a rate of 1.7 per cent during the second period. Well irrigation, on the other hand, shows the opposite trend, growing at a modest 1.6 per cent rate during the first period and accelerating to 4.4 per cent rate during the second period. Given this pace of growth, it is quite clear that well irrigation will increasingly dominate the irrigation picture in India.

Table 1.4 : Percentage Growth and Average Statistics for Net Irrigated Area by Source, 1951–83

Variable	*1951-65*		*1968-83*	
	Growth (Per cent)	*Average (1,000 ha)*	*Growth (Per cent)*	*Average (1,000 ha)*
Total Net Irrigated Area	1.7	23,321	2.4	34,352
Net Area Irrigated by private canals	-0.5	1,241	-3.4	842
Government canals	2.3	8,465	2.3	12,835
All canals	2.0	9,706	2.0	13,677
Wells	1.6	6,941	4.4	14,412
Tanks	2.4	4,312	-1.7	3,837
Others	0.0	2,361	0.4	2,425

Fourthly, regionally, as seen in Table 1.5, expansion of net irrigated area has been uneven. Growth was strongest in a band stretching transversely across the western peninsula of India up through Uttar Pradesh. For all of India, growth was roughly twice as rapid in the areas irrigated by wells as in the canal command, except in the western region where canal irrigation grew most rapidly. In the south , expansion was virtually nil as 'shrinking tank' and other irrigation was compensated for by increased well irrigation. In eastern India, strong growth in the area irrigated by wells was partially offset by the contraction of the area under tank and other sources.

Table 1.5 : Average Net Irrigated Area by Region and Source

*Region and years averaged**	*Canal*			*Tanks*	*Wells*	*Others*	*Total*
	Government	*Private*	*Total*				
Southern	2,660	6	2,666	1,671	1,826	193	6,356
Northern	2,723	179	2,901	3	2,911	119	5,934
Uttar Pradesh	3,206	1	3,207	188	5,698	316	9,410
Central	2,058	1	2,259	303	2,885	220	5,467
Eastern	1,372	512	3,044	608	1,360	1,101	6,113
Western	1,372	25	1,357	634	3,020	283	5,293
All India	14,534	723	15,275	3,407	17,753	2,494	38,912

* Average of 1980-81, 1982-83.

Source : Indian Agriculture in Brief, Department of Agriculture (New Delhi).

In sum, it shows the steady growth in irrigated area as a result of both public investment in canals and largely private investment in well irrigation. They also suggest the general and increasing importance of well irrigation in India since 1965, in particular its extremely strong rate of growth in eastern and central India in recent years.

Different sources of irrigation like canals, wells and tanks were developed at different times as demand for irrigation water as well as technologies for storing, transporting and lifting water developed. Large-scale canal irrigation began to develop only over the past one hundred years, when more massive investments in irrigation were made possible, based initially upon massive economic interests in irrigated agriculture for export goods under British rule (Von Oppen *et al.*, 1983) and later, after Independence, out of the need for sustaining self-sufficiency in food production. The cumulative growth of irrigation potential in India indicates further that canal projects accounted for about 68 million acres or about 48 per cent of the total potential developed (Table 1.6). The fact that major irrigation is essentially a post-Independence phenomenon can be clearly recognised from the investment pattern in the successive Five-Year Plans. A lion's share (78 per cent) of such investment has been on major and medium irrigation projects. But unfortunately, the potentials created under major and medium projects have not been fully utilised. By 1985, about 83 per cent of the potential created has been utilised, leaving 17 per cent unutilised[4]. Even in the case of potential utilized, we are not, in fact, very much aware of the nature and distribution of the effects that this irrigation has upon the economy in general (Stone, 1984). All these issues have been a matter of concern.

Table 1.6 : Estimates of Ultimate Irrigation Potential for Surface Water and Ground water in various years: 1972-88

(Million hect.)

Year	*Source*	*Surface water*	*Ground water*	*Total*
I	*II*	*III*	*IV*	*V*
1972	Irrigation Commission	59.0	22.0	81.0
1975	Fifth Five Year Plan	72.0	35.0	107.0
1985	Seventh Five Year Plan	73.5	40.0	113.5
1988	Planning Commission	98.0	50.0	148.0

There is a view emerging where it is held that canal irrigation has proved a costly experiment. If one accepts this view, then the immediate question is : what are the factors that led to the emergence of such a view ? Till recently, irrigation has been considered synonymous with the construction of dams and canal network. The distribution and management of canal irrigation water has received much less attention. Many irrigation projects have been designed and managed almost exclusively from an engineering point of view (Thavaraj, 1982). The socio-economic and environmental dimensions do not seem to have been integrated with technical design. So, it is sometimes argued that it is not correct to characterise irrigation projects as purely hydraulic system to be run strictly according to engineering principles. All participants—like farmers, irrigation officials and politicians—are expected to play their role to maximise their private interest to the cost of social goods (Caruthers, 1987). It is also, in essence, an integral part of a socio-eocnomic system.

The policies of irrigation water distribution and use appear to have not taken into account the responses and understanding of the cultivators. The need for understanding the nexus between the technology per se and its operationalization has a special significance in a democratic system. Caste, class and factional ties and antagonisms have become relevant in the delivery of any goods and services to the community (Srinivas, 1984). The mismatch, if any, between theoretically assumed policies and goals and actual results in a project leads to a chain of adverse socio-economic and environmental problems.

It has been frequently observed that the investment of scarce capital resources in irrigation projects results in low returns. Delay in completion of projects results in costs escalation. There are projects which are under construction for more than thirty years. Inter-state water disputes often contribute to the spread of available resources thinly on multiple projects. There are many other issues that need to be answered before one raises the issue of cost-effective benefit from major irrigation (Table 1.7).

Table 1.7 : Discrepancy between Original and Revised Cost Estimates in some Irrigation Projects

	Name of the Projects	*Percentage of cost escalation*
Major	Hirakud	230
	Mahanadi Delta	426
	Salandi	360
Medium	Kalo	248
	Salia	592
	Ong	315

Source : Report of the Comptroller and Auditor General of India (1977–78) (Civil Orissa).

The researchers seem to be still at large to identify and analyse the factors contributing to the supposedly unsatisfactory performance of major irrigation projects. The emphasis in canal irrigation has been and continues to be the construction of new projects rather than the management – the operation and maintenance – of the existing systems. Most of the canal systems in India are mainly administered rather than managed (Seckler, 1981).[5]

Under this problems mix, it is pertinent to examine empirically some of the issues and problems of water use and distribution in major irrigation projects, both at the main system as well as tertiary levels mainly to identify the operating constraints. This will help improve the existing projects as well as improve the planning for new projects. Given the problems of higher construction costs and less favourable sites for new projects, the objectives of increasing the irrigated area and agricultural production could be achieved only through improving the existing systems. Many projects have not been improved due to want of adequate research basis to identify appropriately such performance constraints. What is, therefore, needed is a thorough evaluation of the canal irrigation system. This study has been done mainly to fill up this research gap.

NOTES

1. The classification of Irrigation Projects in India is on the basis of extent (size) of Cultivable Command Area (CCA) serviced by an irrigation work. A Project with a CCA of more than 10,000 ha. is called 'Major Irrigation', with a CCA of more than 2,000 ha. but less than 10,000 ha. is called 'Medium Irrigation' and a Project having a CCA upto 2,000 ha. and less is called 'Minor Irrigation'.
2. The corresponding figures for China are 72 per cent and 85 per cent and for Israel. 100 per cent and 99 per cent. More disturbing is the fact that today only 3 per cent of the Indian rural population have access to sanitation as against 81 per cent in China and 95 per cent in Israel.
3. Two important points are in order. In the first place, it should be mentioned that as of 1984–85, about 60 per cent of the ultimate irrigation potential was used. Groundwater use was about 70 per cent and surface water use was about 54 per cent. Only in the eastern and central regions, the unexploited potentials exceeded 50 per cent of the ultimate and the average of unexploited potentials of the entire country was just over 40 per cent. In the eastern region, the unexploited share of groundwater was largest; while, in the eastern and central regions, that of surface water was greatest. In both cases, however, the bulk of the absolute amount of remaining potential lay on the surface, Secondly, estimates of ultimate irrigation potentials do not remain constant. Table 1.6 demonstrates how the value has expanded since 1972, with the total ultimate potential growing by about 75 per cent between 1972 and 1988.

   Of course, this uninterrupted increase in ultimate potential estimates seems unlikely, at least for surface water for some reasons; because, a large portion of the increase in estimates simply represents a refinement based on better techniques and better information. Secondly, a significant part of the increase is supposed to come from major trans-basin diversion schemes. These are expensive and hence sometime in the past proved elusive. And again, as the easily exploitable sites are exhausted and the objectives of irrigation shift from protective to productive, the cost of surface irrigation development would be gradually increasing. In the groundwater sector too, the resource is hidden from view and hence assessment is more difficult and imprecise. In fact, exploitative potential may not continue to increase significantly. See Mark Svendsen in 'Sources of Future Growth in Indian Irrigated Agriculture' in *Future Directions for Indian Irrigation : Research and Policy Issues* (Ed.) by Ruth Meinzer-Dic and Mark Svendsen, IFRRI, July, 1991, pp. 51–53.

4. A recent study estimates that about 75 per cent of the major and medium irrigation systems in India probably utilised only 25 per cent of their potential capacity. Malhotra *et al.* has quoted C.C. Patel as saying : 'We are operating our canals sub-optimal' in S.P. Malhotra *et al. A Performance Monitoring Study of the Warabandi System in Irrigation Management in North West India*, The Ford Foundation, 1983.

5. The unofficial estimates of the total costs of new Medium sized surface irrigation projects are from Rs. 6,000 to Rs. 10,000 per acre in real terms, almost double the cost ten years ago. See David Seckler, *The New Era of Irrigation Management in India*, The Ford Foundation, India, 1981.

   Cost escalation of major and medium projects during 1946–69 on average for India was 108 per cent. Taking Statewise overall figures of the medium projects above, in terms of cost escalation during 1971–81, all-India average was 48.94 per cent. See Niranjan Pant, 'Major and Medium Irrigation Projects : Analysis of Cost Escalation and Delay in Completion' in *Economic and Political Weekly*, Vol. XVII, No. 26, June 26, 1982.

   Table 1.7 presents the percentage increase in costs of construction of a few major and medium projects worked out on the basis of Reports of the Comptroller and Auditor General of India.

TWO

IMPACT OF CANAL IRRIGATION DEVELOPMENT ON THE FARM ECONOMY

Impact of irrigation development on a less developed farm economy can be far-reaching both in dimension and magnitude[1]. There are many economists who indicated the pervasive character of irrigation impacts that operate through interaction of area, yield and crop pattern phenomenon. In the precise context of agricultural production, the direct effects of irrigation can be said to consist of (a) Area or Acreage effect, (b) Crop pattern or crop substitution effect, (c) Yield effect, and (d) Total output effect. Economists like Bharadwaja, (Bharadwaja, 1974) have maintained that irrigation can raise the productivity in three ways, i.e., by making possible multiple cropping, by increasing the yield per unit cost and allowing the production of more lucrative crops. Gadgil (Gadgil, 1948) has indicated that the main direct effect of irrigation is to put previously cultivated lands to new uses or to make them more productive in former uses or to bring new lands under use for the first time. Similarly, Singh (Singh, 1979) has shown that the provision of irrigation causes higher crop yields per unit of land and generates increased agricultural income.

An impact of irrigation through which it increases the gross sown area is called its area effect. It has two components : in the first place, increase in net sown area which becomes possible on account of facilitating reclamation of waste land and extending cultivation to hitherto uncultivated land; secondly, increase in gross cropped area arising out of irrigation encouraging intensive cultivation of currently cultivated land, that is, the index of cropping intensity. Both these through positive interaction lead to absolute increase in gross cropped area.

Another impact of irrigation on an agricultural economy is the changing crop pattern. Of course, all changes in cropping pattern cannot be said to be irrigation-induced. Some changes in cropping pattern do take place on account of economic and institutional factors, such as urbanization, changes in food habits, land tenurial conditions, relative price movements of agricultural products and state policies on subsidies.

It is also true that irrigation enhances the yield rate per hectare as it enables the farm operations to switch over to improved technology. This change in the character of cultivation under irrigation leads to greater productivity of crops previously cultivated. Ample evidences suggest that yield on irrigated land is markedly higher than that on unirrigated land. Some estimates which are admittedly approximate suggest that across states, gross value of output per hectare of irrigated land is between 50 and 280 per cent higher than that of unirrigated land.

Table 2.1 : Productivity Differences between Irrigated and Unirrigated Land

State	*Absolute (Rs. per hectare)*	*Ratio*
Andhra Pradesh	1,236	3.3
Assam	n.a.	n.a.
Bihar	600	1.5
Gujarat	1,250	2.8
Haryana	1,355	3.6
Jammu and Kashmir	n.a.	n.a.
Karnataka	1,265	3.2
Kerala	n.a.	n.a.
Madhya Pradesh	605	1.9
Maharashtra	1,325	3.8
Orissa	389	1.7
Punjab	1,465	2.5
Rajasthan	900	3.5
Tamil Nadu	1,677	3.6
Uttar Pradesh	980	1.7
West Bengal	n.a.	n.a.

Sources : (I) A. Vaidyanathan, 'Irrigation and Agricultural Growth', in *Indian Journal of Agricultural Economics,* Oct.-Dec. 1987.
(II) Centre for Monitoring Indian Economy, Basic Statistics relating to the *Indian Economy*, Vol. 2 (New Delhi, CMIE,1988).

Differentials of a much larger order are apparent when we consider territorial units smaller than the states. Moreover, except where irrigation is largely from small storage works that depend mainly on local rainfall, the productivity of irrigated land appears to be more stable across years. Again it is also observed that such changes in per acre yield are technology-induced; and irrigation constitutes the major causative factor behind such yield rate enhancement.

It has been well borne out by a number of studies that availability of irrigation facility leads to enhancement in mandays of employment in farm operations. Irrigation operation itself needs more labour. Its availability generates on-farm employment through increased cropping intensity and adoption of improved and labour-intensive cultural practices. It also generates off-farm employment through larger volume of transactions of agricultural goods and their processing. In an abundantly labour-surplus rural economy, irrigation reduces the rigour of unemployment, under-employment and seasonal unemployment. Of course, there are conflicting views. Some economists argue that there has not been much of a significant expansion of employment.[2]

Irrigation, with its character of high potential and scale neutrality, has also offered a scope for the 'disadvantaged' group in respect of derivation of benefits from irrigation. While investments in infrastructure and production have been on the rise, there is a need for a better percolation of the development benefits to the less endowed areas and the weaker sections of the producing community. It may be true that irrigation (particularly canal irrigation) is scale neutral and as such, there may not exist any inherent tendency to make existing income distribution more unequal. Of course, there are some evidences to show that inverse relationship between size of farm and production has somewhat weakened in recent years. Economists such as Dandekar, Minhas and Sau have expressed apprehension that irrigation would in its turn widen the existing disparity in the interpersonal distribution of income. It is fair to say that for an explanation of this trend, the operational features of the irrigation system, particularly canal irrigation system and other technology, need to be examined against the backdrop of the prevailing structural and organisational bases of individual farms. The roots of the problem may be embedded in a complex nexus between the irrigation system and other technological inputs, farm level resource endowment and the type of institutional supports available. Big farmers may have an edge over small and marginal farmers. It is important to see whether in an environment devoid of limitations in respect of inadequate resource base and insufficient infrastructural arrangements surrounding them, one may visualize canal irrigation living up to its virtue of neutrality. It is a verifiable issue whether regressive impact on income distribution of canal irrigation may be caused by the non-neutrality of economic, social and political institutions rather than the new technology, canal irrigation, *per se*.

An Illustrative Micro-Study

To assess the economic consequences of provision of public irrigation facility, a micro-level[3] field study was conducted in a homogeneous region in the command area of the Mayurakshi Canal Irrigation Project[4] (for Mayurakshi Canal Irrigation Project, see Annexure 1).

An illustrative micro-study on irrigation impact involving a homogeneous tract is considered necessary. Variation in soil types, agro-climatic conditions, inflow of resources, availability of farm labour and infrastructural facilities as between different areas of the State and even within an area emphasises relevance of such a study in a disaggregative framework. Evaluation of the different aspects of the relative profitability of irrigated farming over that of unirrigated farming would also help evolve suitable policy decisions with respect to expansion of irrigation and extension of associated farm inputs. Specifically, this micro-study would attempt to estimate the incremental farm business income and mandays on farm employment resulting from provision of irrigation facilities. Again, the issue of distribution of irrigation benefits as between different size-classes of farm operators can be analysed. Another important aspect in the context of irrigation benefits is that provision of irrigation results in increased crop production as well as increased use of farm inputs, but how much of the increased production is attributable to the irrigation factor alone is to be estimated by fitting an appropriate production function using the data from the micro-study (for Production Function Analysis, see Annexure 2).

Objectives

Attempts have been made to examine, among others, the following hypotheses :

Hypothesis I : Provision of canal irrigation leads to substitution of less profitable crops like jowar, bajra, maize etc. by more lucrative crops like potato, mustard oilseeds, sugarcane, vegetables etc. Availability of canal irrigation water has resulted in diversification of cropping pattern. Because of irrigation, crop intensity in respect of each of the size-groups of holdings has been increased.

Hypothesis II : The employment level in the irrigated tract is equal to that for unirrigated zone.

Hypothesis III : Assuming a linear relationship between inequality (X) and irrigation (I),

$X = a + bI$

Null Hypothesis : $b=0$

Alternative Hypothesis : $b \neq 0$

Benefits from irrigation (canal) are not confined to operators of large farms, but are enjoyed by farmers of all size-groups.

The Study Area

As a prelude to the main analysis, some aspects of the study area, methodology, structure of the farms and their organisations, the sample etc. are in order.

Agriculture is the mainstay of the people of the district of Birbhum. According to Census, 1991, the total number of workers in the district was 6.9 lakhs constituting 34 per cent of the total population of the district. Of these workers, again, the cultivators and agricultural labourers account for 16 per cent and 11 per cent of the total population respectively. About 74 per cent of the total working population of the district is engaged in agriculture as against 54 per cent in the State of West Bengal. The total cultivable area of the district is 9,69,000 acres. The above facts indicate the pronounced agrarian character of the district.

In this district of Birbhum, having for the most part, a porous land and low water holding capacity and being almost wholly (about 80 per cent) aman growing area which needs water for its very existence from the time it comes out as a seedling till it attains its maturity, demand for irrigation water, especially at a time when there is insufficient or erratic distribution of rainfall (monsoon rainfall does not follow a rigid pattern), is considered always very urgent. Monsoon remained the traditional dispenser of the district's destiny.

The Mayurakshi canal could inject life into agriculture of this district which had a sad record of periodic crop failure. The canal project has a command area of 853 square miles within the district. The net irrigation area under Mayurakshi canal command is about four lakh acres of land. This canal is a typical irrigation project launched in conditions highly favourable to its success and with an assured project of solid benefits accruing from it.

It is evident from Table 2.2 that Mayurakshi Canal Irrigation Project is the predominant source of irrigation in all the blocks. The canal's share in total irrigation area varied between 85 per cent to 100 per cent except blocks like Sainthia and Bolpur in Birbhum district (the shares of the area irrigated by canal source to total irrigated area being 28 per cent and 68 per cent respectively). The other sources of irrigation in these areas in order of importance are shallow tube-wells and tanks.

Table 2.2 : Irrigated Area by Sources of Irrigation : District Birbhum

Blocks	*MC I*	*Tank*	*Tube-well*	*Others*	*Total*	*Percentage of canal irrigated area to total irrigated area*
I	II	III	IV	V	VI	VII
Murarai	402	16	—	—	418	96
Nalhati	789	66	—	—	856	92
Mayureswar	2,225	159	4.0	—	2,389	93
Rampurhat	867	38	—	—	906	95
Md. Bazar	445	79	—	—	523	85
Sainthia	324	10	8.27	—	1,162	27
Suri	364	—	—	—	364	100
Dubrajpur	161	10	—	—	171	94
Illambazar	52	—	—	—	52	100
Bolpur	906	68	—	87	1,061	85
Labpur	145	59	10	—	204	67
Nanoor	963	—	—	—	963	100
Total	7,643	505	13.42	87.00	9,069	84

Source : Survey Report on Mayurakshi Canal Irrigation Project, 1996, by M.G. Ghosh, p .27

The Methodology

Normally, two types of methodology are generally being used by the research workers for explanation of problems like evaluation of benefits from any canal irrigation projects. They are : (a) cross-section comparison of irrigated and non-irrigated farms ('with' and 'without' principle); and (b) comparison of changes before and after irrigation ('Pre-Project' and 'Post-Project' principle). 'With' and 'without' principle can evaluate irrigation impact in terms of the difference it makes, that is, in terms of the effects which it specially causes.

Both these principles have their merits and demerits. In the case of 'pre-project' and 'post-project' analysis, it is essential to take due note of some factors which change along with irrigation. The intervening period is one of price change. The reported change in income of sample farmers needs to be duly adjusted. All input output data are to be readjusted in constant prices. Again, weather conditions, 'before' and 'after' irrigation, may change and as such, situations may be far from comparable. So, results, in respect of impact, may be considerably distorted, both in direction and magnitude. Furthermore, in agriculture, new varieties of crops may come up to be introduced after the introduction of irrigation. This may generate the possible distortion on irrigation impact. This distortion may, however, be eliminated to the extent their adoption is not contingent on having access to irrigation. Another disadvantage of the 'pre' and 'post' approach lies in that information relating to the pre-project period depends largely on the ability of the respondents to recollect from memory. On the other hand, the 'with' and 'without' principle duly eliminates the influence of change other than that due to irrigation. So, there is no problem of price adjustment to be tackled, nor has one to worry unduly about the problem of weather changes. Of course, the approach may be perfectly effective if control sample does not differ from the main sample in respect of soil, topography, farm size and management ability etc. It is difficult to identify such a situation. Another limitation is related to the courageous assumption that situation – 'without' irrigation – will be a case of total dry farming –a situation which sometimes may be conspicuous by its absence. Despite these limitations, we have in our study adopted the 'with' and 'without' principle. This methodology is to be used mainly because of the absence of any 'benchmark' survey relating to the Mayurakshi Canal Irrigation Project.

By and large, the concepts of 'Cost' and 'Returns' used in the analysis are the same as those generally adopted in Farm Management Studies conducted in the country.

Gross Revenue (GR) is defined as the total physical output, that is the main product and the by-products of farms valued at farm harvest prices. Another concept used is that of Net Revenue (NR). This is a strict accountancy view of farm business in that it is derived by deducting from gross revenue all costs – out of pocket as well as imputed –incurred in the process of cultivation. The concept of Farm Business Income (FBI) has also been used. This may be defined as the gross revenue minus all out of pocket expenses incurred in cultivation. FBI is a highly relevant variable which the farmers normally attempt to maximize. The cost concepts used here are as follows :

Cost A_1 : Cost of hired human labour, farm and hired draft animal labour, seed and manure, both purchased and farm grown, fertilizers, insecticides, interest on circulating capital, land revenue, irrigation charges, etc.

Cost A_2 : Cost A_1 plus rent paid on leased-in land.

Cost B : Cost A_2 plus interest on capital invested.

Cost C : Cost B plus imputed value of human labour added by the farm family.

In our study, the costs of hired labour have been evaluated at wages prevalent in different seasons for agricultural year. The cost of family labour has been evaluated at the wage rate of the casually hired labour. The working cost of owned bullock labour is equal to the prevalent rate for hiring a pair of bullocks, in view of the difficulties involved in correctly ascertaining the maintenance cost of own bullocks by interview method.

While assessing the value of seeds, the prices of different seeds prevalent in the village at the time of sowing have been taken into account. The value of fertilizers and pesticides has been taken according to their market value. Depreciation on fixed capital has been charged on the basis of straight line depreciation method. Irrigation charges and land revenues have been calculated on the basis of actual payments made. The rental value of own land was calculated as equal to 20 per cent of the total value of the produce obtained from all crops during the year. This approach , of course, is in conformity with the latest available farm management procedure. Interest on fixed capital has been charged at 10 per cent per annum. Interest on working capital has been fixed at a higher rate of 12 per cent per annum.

The period for which the data were collected refers to the 1995-96 agricultural year i.e. Kharif 1995 and Rabi 1996. The data were collected by the researcher himself. In this connection, it should be noted that all efforts were made to elicit as accurate information as possible.

The Structure of the Farm and the Sample

Since the study pertains to irrigation, Birbhum district, having good canal irrigation incidence because of Mayurakshi canal irrigation, was purposively chosen.

With regard to the sampling technique adopted for field investigations, it was based on two-stage stratified random sampling with operational holding as the ultimate unit of sample. At the first stage, six villages (three irrigated and three unirrigated) were selected for study. For the selection of the sample unit at the second stage, a list of farming families was prepared in each selected village and the particulars of their land holdings etc. were verified with reference to records available at the village and block levels. Then farming families were divided into three strata as per the size of operational holdings of '0' to '0.9' hectare (size group I), 1 to 1.9 hectares (size group II) and 2 hectares and above (size group III).

From each stratum, 40 per cent sample households were selected at random. Total households thus selected from irrigated and unirrigated villages were 120 and 110 respectively. The village-wise (or overall) distribution of 230 samples has been presented in Table 2.3

Table 2.3 : Villagewise Distribution of Samples under Different Size Groups in Irrigated and Unirrigated Areas

	Number of Samples drawn			
	Size-Class I	*Size-Class II*	*Size-Class III*	*Total IV*
Irrigated Villages : Overall	62	42	16	120
Unirrigated Villages : Overall	51	40	19	110

As regards different aspects of the farm economy and comparison between irrigated and unirrigated samples, the following observations may be made. As it is evident from Table 2.4, the land holdings in the sample farms are skewed to the left under the irrigated as well as unirrigated conditions. For example, in irrigated area, 86.70 per cent of the farm households belonging to size classes I and II together control 65 per cent of the total net sown area. Similar pattern of land holdings also prevails in case of unirrigated farms, where 85 per cent of farm households of size classes I and II together own 63.11 per cent of the total net sown area. This distribution pattern is, in fact, not much different from that of the tract as a whole under study.

Table 2.4 : Distribution of Sample Farms by Size-Class

(a) Irrigated Area (in hectares)

	Size-Class	*I*	*II*	*III*	*Total*
I.	Number of holdings	62.0	42.0	16.0	120.0
II.	Holdings in the size-class as percentage to all holdings in sample	51.7	35.0	13.0	1.C
III.	Aggregate area in size-class	35.9	49.0	47.0	131.9
IV.	Aggregate area in size-class as percentage of total area	27.5	37.5	35.0	100.0
V.	Mean size of holdings	0.58	1.16	2.93	1.2

Contd.

	Size-Class	I	II	III	Total
(b) Unirrigated Area (in hectares)					
I.	Number of holdings	51.0	40.0	19.0	110.0
II.	Holdings in the size-class as percentage to all holdings in sample	46.36	36.36	17.28	100.00
III.	Aggregate area in size-class	34.0	59.0	52.0	145.0
IV.	Aggregate area in size-class as percentage of total area	23.43	40.68	35.89	100.0
V.	Mean size of holdings	0.66	1.47	2.73	1.31

On Cropping Intensity and Cropping Pattern

The gross cropped area, net sown area and the index of crop intensity in irrigated and unirrigated areas are presented in Table 2.5.

Table 2.5 : Gross Cropped Area (GCA), Net Area Sown (NAS) and Index of Crop Intensity (ICI) in the Study Area

Size-class	*G.C.A.*	*N.A.S.*	*I.C.I.*
I	*II*	*III*	*IV*
(a) Irrigated Area	Hectares	Hectares	Hectares
I	49.00	25.41	154.00
II	69.00	43.00	168.00
III	71.80	52.00	149.00
Overall	189.80	120.41	160.50
(b) Unirrigated Area			
I	40.50	32.00	125.00
II	51.00	40.50	122.00
III	62.00	47.00	119.00
Overall	153.50	119.50	129.00

It is evident that the index of overall cropping intensity in the irrigated area is 160.5 as against 129.00 in the unirrigated area. It is also found, size-class-wise, that in the irrigated area, the medium size-class farmers have the highest crop intensity of 168.00 and the highest size-class farmers the lowest crop intensity of 149.00. In the unirrigated area, the small size-class farmers have the highest crop intensity of 125.00 and the third size-class farmers have the lowest intensity of 119.00.

It can also be concluded that the acreage effect of irrigation, i.e., increase in gross cropped area resulting from enhanced cropping intensity is 29.50 per cent. Our hypothesis that provision of canal irrigation here results in increased cropping intensity is confirmed by the findings of the case study.

It is also important to mention that the causes as to why the crop intensity index in the irrigated study area has increased only to 160.50 are many and varied. The absence of water management practices, particularly that of field channels was the major constraint in achieving a high cropping intensity. Again, the empirical evidence gathered indicates that utilisation percentage is the highest in the head reaches than in the tail-end areas. Canal irrigation at the tail-end areas would have to make do with less water using because of water not being supplied to them adequately. The tail-end areas are often left without water. So, tail-end farmers are reluctant to grow HYV paddy and vegetables; summer crops and pulses are not even grown. So, further increase in crop intensity is retarded. In the unirrigated areas, the small size-class farmers have the highest crop intensity of 125.00 and the big size-class farmers have the lowest intensity of 119.00 per cent. Raising two full crops on a plot in unirrigated zone during a year would be an exception rather than a rule.

Irrigation segment in our analysis is represented by a wide array of crops. It has been observed that paddy is grown in 100 per cent of the net area under both irrigated and unirrigated conditions. The areas under Kharif paddy has remained more or less unchanged; while the area under boro paddy has shown a remarkable change during the period. The area under boro paddy was only 200 hectares in the early 50's which has increased to 33,900 hectares in the late 80's. Boro (Summer Paddy – high yielding variety) paddy, a heavily-irrigated crop, is not significantly grown in the unirrigated regions and tail-end regions of the canals, predominantly because of erratic and uncertain supply of canal water. Only 49 and 51 per cent of the net sown area located in the head reach and middle reach areas respectively of the irrigated zone are allocated for the summer boro paddy.[5]

In the case of irrigated farms, pulses, potato, mustard oilseeds and vegetables are grown in 11 per cent, 6 per cent, 4.8 per cent and 3.4 per cent respectively in the head reach areas; 18 per cent, 5 per cent, 6 per cent and 2 per cent respectively in the middle reach areas; and 2 per cent, 1 per cent, 2 per cent and 0.4 per cent respectively in the tail-end areas of the irrigated tract. No significant areas for the cultivation of the above-mentioned crops have been allocated in the unirrigated zone. Only pulses are grown in as much as 16 per cent of the total net sown area in the rabi season. Pulses can withstand relatively low and uncertain irrigation water. Concentration of heavy water using crops like sugarcane has been found on head reaches of the canal. Of course, it does not account for a large percentage of net irrigated area.

The overall cropping pattern in the study area, therefore, broadly indicates that the availability of irrigation water has resulted in diversification of cropping pattern. It is also pertinent to see that in the unirrigated farms, low priced pulse crops, paddy etc. have been generally grown extensively; whereas high value crops like HYV paddy, summer vegetables etc. are now grown under irrigated tracts. Even though pulses involve low cost of production as they could be raised under rainfed and retained soil moisture conditions, the cultivators do not consider it at all a paying proposition, notwithstanding their low cost of production. So, it emerges that provision of water leads to a substitution of less profitable crops by more profitable lucrative crops. The change in the choice of crop pattern is largely due to the transformation of agriculture from subsistence-oriented towards market-oriented changes in labour relations, changes in practice of payment of wages in cash rather than in kind etc. Those farmers are found to be more progressive, who use fertilizer, introduce new crops, use more hired labour and have, in general, attuned farm management more to market demand. Based on the above observations, one is invariably led to the conclusion that there has been a departure from the traditional subsistence pattern to a market-oriented commercialized crop pattern.

Another important observation that emerges in this connection is that in the study of crop pattern in relation to size group, no remarkable divergence in the cropping pattern has been noticed. Of course, only in the rabi season, in the irrigated area, the farmers in the lowest size group of holding have allocated only 38 per cent of the area for high yielding variety paddy, 1.8 per cent of the area for vegetables and 6.5 per cent of the area with pulses. This is quite natural, because HYV paddy and vegetables are water- intensive and labour-intensive crops which the small farmers cannot generally afford. On the other hand, the relatively big and medium farmers who largely depend on hired labour for farm operations often tend to devote proportionately larger area for raising these water-intensive, labour-intensive crops. In the unirrigated area, practically no divergence in the cropping pattern with the changes in the size group, both in the kharif and rabi seasons, has been generally observed.

It also emerged that the impact of canal irrigation on cropping pattern is not uniform in all the localized areas (Head Reach–HR, Middle Reach–MR, and Tail end–TE). The study reveals that the canal water is not evenly diffused in all the areas. The performance of farmers at the tail-end is almost nearer to farmers in the unirrigated areas; whereas, for all practical purposes, farmers in middle reach and head reach areas fare better. They go in for heavily-irrigated crops like paddy, potato, vegetables etc. which require high volume steady and regular supply of water and so, more dominant. They share much more water for their crops which create scarcity and hence non-availability of irrigation water at the tail-end of the command area. This kind of development also sometimes results in waterlogging, soil alkalinity and salinity in the upper reaches.

On Employment

It is true that with the traditional agriculture involving low index of crop intensity and monocrop pattern under unirrigated farming practices, there appears to exist considerable under-employment, seasonal unemployment and unemployment in the rural sector, where cultivators and agricultural labourers constitute about 70 per cent of the labour force. This problem is further aggravated by low pace of industrialization. The main thrust of absorption of the surplus rural manpower lies in the agricultural sector itself. In view of the protective, productive and employment generating capacity of irrigation, there is a need for its proper development.

The number of mandays of employment per hectare of net sown area per annum for all crop enterprises in irrigated and unirrigated farming conditions have been worked out and presented in Table 2.6. Two important points are to be noted here. In the first place, the mandays of employment refer to own-and-family labour as well as hired labour. Secondly, the hired labour in the sample areas is of three categories:

(a) Casual hired labour consisting of male and female workers employed on daily wage basis;
(b) committed labour attached to the employer for a specific period, say one year, who is employed on daily wage basis; and
(c) permanently attached labour as employed on an annual basis with board, lodging, some clothing and cash payment. They are normally known as farm servants usually hired by big land holders.

Table 2.6 : Mandays of Employment under Irrigated and Unirrigated farm conditions

(per acre of net sown area- for all crops basis)

Size Class	*Average mandays of employment (unirrigated farms)*	*Average mandays of employment (irrigated farms)*	*Net increase in mandays of employment*	*percentage change in employment*
I	II	III	IV	V
I	86	161	75	185
II	92	164	72	177
III	87	160	73	174
Overall	85	162	77	195

Source: Computed from the data in the ICSSR Survey Report, 1995.

The Table 2.6 shows that on an average, the irrigated farms engage 162 mandays as against 85 mandays in unirrigated farms per acre per annum. In other words, an acre of irrigated farm increases the labour use by 195 per cent. So, canal irrigation has pushed up the labour employment opportunities in agricultural sector.

Variations in labour use per cropped area may be seen as the combined effects of variations in proximate factors like higher intensity of cropping, diversification of crop pattern, cultivation of labour-intensive crops etc. Ploughing, sowing, transplantation, weeding, harvesting–all are important labour-intensive operations in crop production. A small percentage of labour is used generally for manual work – presumably, supervision.

The complementarity between irrigation and level of use of biochemical technology has important bearing on the intensity of the use of labour. Another important observation is that the cultivation of summer paddy has helped to remove largely seasonal unemployment of rural labour force. In fact, there are only two peak periods of employment – transplantation and harvesting of aman rice. But now these days, there are four peak periods: two peak periods each for aman rice and summer boro crops. To meet the increasing demand for agricultural labour during the peak seasons, a large number of migrant labour from the adjoining districts use to come, more often than not, to work in the field. Clearly, it may be concluded that the introduction of canal irrigation followed by the adoption of HYV paddy was found instrumental in generating substantial amount of employment opportunities. This hypothesis is also positively affirmed by our own findings from the micro-study.

On Costs and Returns

The estimates of different costs, gross value of output and indices of farm profit such as farm business income, returns to family labour and management and net returns have been worked out for total farm enterprises per hectare for an agricultural year to throw light on the issues of relative efficiency and income distribution pattern of different farms with and without irrigation. The results of field survey on costs and returns are shown in Table 2.7. The Table gives the average costs of cultivation (cost A_2, cost B and cost C) and Farm Business Income (FBI) and Net Returns (NR) of net sown area of the different farm groups under irrigated and unirrigated farm conditions.

Table 2.7 : Different Costs of Cultivation and Returns on the Sample Farms (Rs. per acre of net sown area)

(Incremental)

Size-class	*Cost A_2*	*Cost B*	*Cost C*	*FBI*	*NR*
(a) Irrigation Area					
I	510	640	710	820	362
II	540	685	704	640	220
III	548	700	705	501	271
Overall	485	675	695	590	245
(b) Unirrigated Area					
I	301	315	314	458	—
II	295	310	320	420	—
III	258	301	319	230	—
Overall	298	312	316	345	—

Source : Computed from the data in the ICSSR Survey Report, 1995

It is evident that overall operational cost i.e. A_2 per acre of net sown area in the irrigated farms is higher being Rs. 187 as compared to that of unirrigated farms. Again, the overall cost B in irrigated area is Rs. 675 per acre as compared to Rs. 312 in unirrigated area, the excess being Rs. 363. Furthermore, the overall cost C in irrigated farms is Rs. 695 per acre and in unirrigated farms Rs. 316, the excess being Rs. 379. So, cost A_2, cost B and cost C are consistently higher in irrigated farms compared to the unirrigated area of the sample farm households. It is also clear that cost A_2 consistently increases with the increase in size group of holdings in case of both irrigated farms. Cost A_2 is found to be lower in small-size groups, because of their proportionately higher contribution of family labour and employment of less of hired labour. Large farms relatively use more hired labour. Similarly, with the increase in farm size, there exists an identical trend of increase in cost B. Of course, cost C shows a different trend as compared to cost A_2 and cost B. Cost C is highest being Rs. 710 with the lowest size class and decreases gradually with the increase in size of farms. This is because the first size class (small farmers) would have incurred the maximum of costs per acre in irrigated and unirrigated farms, due to absence of contribution of family labour in the case of unirrigated farms. Cost C does not show any consistent trend.

It is the farm profit which is the appropriate indicator of economic efficiency. For this, we have estimated the Farm Business Income (FBI) only separately with respect to irrigated and unirrigated farms size-class-wise. Farm Business Income bears inverse relationship with the size class of holdings in the irrigated and unirrigated areas. The additional Farm Business Income generated by irrigated farms over those of unirrigated farms has been worked out. On an average, the Farm Business Income per acre per annum has increased by Rs. 245 predominantly because of provision of irrigation water. The hypothesis that it is economically more profitable to cultivate on irrigated land stands confirmed.[6]

On Income Distribution

There is no denying the fact that the less developed countries, especially in Asia, since 1950s have invested enormous amount of scarce domestic and international donor agencies, resources for the development of irrigation with the twin objectives of increasing agricultural production and reducing poverty and economic inequality. Irrigation is, indeed, a potent instrument for adding to a farmer's income from crop enterprise. Being land-augmenting in nature, it strengthens the land resource, the main base for income generation in the countryside where land is scarce and skewedly distributed in favour of large farmers, with substantial numbers owning no farm land at all. Advent of irrigation in such a situation might make a difference to the prevailing inequalities of income and wealth.

The informed views on the effect of canal irrigation on income distribution range across a wide spectrum. At one end of the spectrum is the claim that the benefits of irrigation have not been shared equally among different groups of holdings in the same area. Large farmers appropriate the scarce irrigation resources more than proportionately to their share in land resource. Introduction of modern agricultural technology (like irrigation) often creates dualism in the agricultural sector rather than modernising traditional agriculture. In contrast to this, it is maintained that without irrigation and other modern agricultural technology, agricultural technology would have remained below the subsistence level for a longer time. In view of these conflicting claims, it is tempting to appeal to empiricism to settle this issue that the benefits from irrigation are not confined to operation of large farms, but are enjoyed by farmers of all size groups and the uneven distribution of gains involves policy options unconnected with irrigation.

It is against this backdrop that an endeavour has been made to examine the hypothesis: the benefits of canal irrigation are largely confined to upper strata of farm families.

The test of the hypothesis is contingent upon the details of primary statistics on input-output data. The data have been collected in a field survey in the district of Birbhum of West Bengal. The sample households have been selected in a random manner by employing two-stage sampling design. Two villages from each block have been drawn at random by employing systematic sampling technique, giving a number of ten villages among five blocks. Of all the sampled villages, agricultural households have been divided into the following three categories according to the size of net operational holdings, keeping in view the proportion of area and number of operational holdings to the total operational holdings within three categories in the canal command area:

(a) Marginal Farmers: 0-2.5 acres

(b) Small Farmers: 2.5-5.0 acres

(c) Medium Farmers: 5.01 acres and above.

From each size category, a sizeable sample has been drawn. The farmers have been classified into two groups – irrigated and non-irrigated. In all, 200 farms –120 irrigated and 80 non-irrigated farms– have been studied. The information have been collected for the agricultural year 1995-96.

Results and Discussion

Table 2.8 shows that on identical acreage categories, the mean values of all income concepts for irrigated farmers are significantly higher than those for non-irrigated farmers and the percentage change varies inversely with the size category. Irrigated farmers cultivating upto 2.5 acres of land are able to earn 246.45 per cent more farm business income than their non-irrigated counterparts. For the remaining two categories, the estimates stand at 171.34 per cent and 129.44 per cent only. In general, the higher the size of operational holding, the lower are the differential income gains. This finding is corroborated by the behaviour of all the three income concepts employed in the present study.

The antecedent description is confined to household unit gains However, given the heterogeneity in average size of holding, explanation of per acre gains becomes important.

It would be seen from Table 2.9 that per acre income and output gains follow a pattern similar to that of average household gains. The percentage change in per acre gross output, net value added and farm business income registered by irrigated farmers over their non-irrigated counterparts is inversely associated with the size-category.

Table 2.8 : Income Levels of Irrigated and Non-Irrigated Cultivator Households

(mean value for households)

Cultivator Category	*Size Category*			*Overall*
Income level	*A*	*B*	*C*	
A. Gross Value of Output				
Irrigated	10437.77	26643.41	46294.23	21402.54
Non-Irrigated	3976.90	10443.56	20782.87	9242.31
All Cultivators	7938.75	20356.91	34955.85	16575.35
(P_cG)	162.45	155.11	122.75	131.57
B. Net Value Added				
Irrigated	8144.07	21908.68	37713.46	17316.58
Non-Irrigated	2538.90	8452.34	16884.36	7156.73
All Cultivators	5976.06	16866.82	28456.08	13281.91
(P_cN)	220.76	159.20	123.76	141.96
C. Farm Business Income				
Irrigated	7904.07	21548.19	36798.46	16929.45
Non-Irrigated	2281.42	7941.30	16036.31	6705.94
All Cultivators	5729.29	16287.91	27568.06	12869.39
(P_cF)	246.45	171.34	129.44	152.45

PcG : Percentage change in gross value of output.
PcN : Percentage change in net value added.
PcF : Percentage change in the Farm Business Income.

Table 2.9 : Income Levels of Irrigated and Non -Irrigated Cultivator Households

(Rupees Per Acre)

Cultivator Category	*Size Category*			*Overall*
Income level	*A*	*B*	*C*	
A. *Gross Value of Output*				
Irrigated	4213.75	4150.85	4117.42	4154.87
Non-Irrigated	2618.90	2722.40	3118.79	2855.75
All Cultivators	3769.01	3758.26	3796.22	3774.59
(P_cG)	60.89	52.47	32.02	45.49
B. *Net Value Added*				
Irrigated	3287.77	3413.21	3354.24	3361.66
Non-Irrigated	1671.99	2208.33	2533.76	3361.66
All Cultivators	2837.20	3080.70	3090.34	2024.94

Contd.

Cultivator Category *Income level*	*A*	*Size Category* *B*	*C*	*Overall*
(P_CN)	96.64	54.91	32.38	52.02
C. *Farm Business Income*				
Irrigated	3190.89	3357.05	3272.42	3286.71
Non-Irrigated	1502.38	2070.12	2406.49	2072.04
All Cultivators	2720.04	3003.36	2993.90	2931.01
(P_CF)	112.38	62.16	35.98	58.61

To conclude, the gains from canal irrigation are found to be size-neutral. There has been no evidence to suggest that income and output gains are confined to relatively big farmers alone. However, per acre and mean household income gains are more on relatively smaller farmers. In consonance with the above observations, we reject the just hypothesis and conclude that the benefits of irrigation are not confined to upper stratum of farm families alone.

It is to be further noted that despite a distinct increase in the income of each size category, the difference between the household income of the lowest and the highest farm categories is very wide on both the types of farms. From Table 2.9, it is clear that an average farm business income of highest and lowest irrigated farmers gives a range of Rs. 28,889.39. The relevant estimate for non-irrigated farmers stands at Rs.13,754.89. Gross value of output and net value added too assume similar pattern.

However, range is an unsatisfactory measure of dispersion as it ignores the distribution in between the extremes.[7] A better method provided by Lorenz Curve (which takes into account the income of all units as well as their distribution) shows that bottom 19.84 per cent of irrigated households share only 5.54 per cent of farm business income, 5.60 per cent of net value added and 6.06 per cent of gross value of output; while top 19.84 per cent irrigated households share 41.90 per cent of farm business income, 41.79 per cent of net value added and 41.86 per cent of gross value of output. This uneven distribution is also borne out by the fact that the lowest 50 per cent of irrigated households get less income share than that obtained by even the top 10.32 per cent of irrigated households.

Conversely, bottom 20.49 per cent non-irrigated households get 3.70 per cent of farm business income, 4.26 per cent of net value added and 5.83 per cent of gross value of output; while top 20.48 per cent non-irrigated households appropriate 40.48 per cent of farm business income, 48.20 per cent of net value added and 45.40 per cent of gross value of output. Again, the lowest 50 per cent non-irrigated farmers get a share of income less than that obtained by even the top 9.64 per cent non-irrigated households (see Table 2.10).

The Table 2.10 shows large income inequality in both the types of farmers. With respect to Farm Business Income (FBI), Gini concentration ratios stand at 0.373 and 0.443 for irrigated and non-irrigated farmers respectively. The ratios of net value added (NVA) and gross value of output stand at 0.366 and 0.361 for irrigated and 0.434 and 0.387 for non-irrigated farmers respectively.

Table 2.10 : Distribution of Farm Business Income, Net Value Added and Gross Output

(Cumulative Percentage)

Irrigated				Non-Irrigated			
% of household	*% of FBI*	*% of NVA*	*% of gross value of output*	*% of house-hold*	*% of FBI*	*% of NVA*	*% of gross value of output*
I	II	III	IV	V	VI	VII	VIII
10.32	2.37	2.45	2.77	10.84	1.65	1.90	2.57
19.84	5.54	5.60	6.06	20.48	3.07	4.76	5.83
30.16	10.04	10.11	10.64	30.12	6.69	7.42	9.92
39.68	15.10	15.23	16.00	39.67	10.67	11.41	14.65
50.00	22.10	22.27	23.20	50.60	17.40	17.97	21.86
59.52	30.91	31.03	31.89	60.24	25.94	26.60	30.35
69.84	43.13	43.27	44.08	69.88	37.38	37.38	41.28
80.16	57.10	58.21	58.54	79.52	51.52	51.72	54.51
89.68	74.95	74.97	75.06	90.36	74.15	74.51	76.04
100.00	100.00	100.00	100.00	100.00	100.00	100.00	100.00
Gini concentration ratio	0.3731	0.3661	0.361		0.4431	0.4342	0.3877

A perusal of all these estimates leads to the conclusion that the propagation and diffusion of canal irrigation system among agricultural households seem to have reduced rather than aggravated income disparities in the study area. The on-farm benefits from a unit of irrigated area need not rise very much with the size of farm holding. In other words, small farmers can gain, acre for acre, as much benefits from irrigation as do large farmers. This is borne out by the survey. This leads us to the falsification of the hypothesis that irrigation has increased the uneven distribution of income among farm families.

NOTES

1. Ishikawa has quite rightly considered irrigation as the 'leading input' as much as it is a precondition for other land-augmenting changes (such as higher cropping intensity, more intensive use of fertilizer, etc.) to take root. see Ishikawa, S. (1967).
2. It may be of interest to find out through a micro-level study like this how the project has fared in the context of the issues raised here in an area of West Bengal. Micro studies as these by their nature cannot be blown up for any overall estimation. But the plus point of such problem-oriented micro-studies is that one has an opportunity to have a close view of the problem.
3. In point of fact, Mayurakshi canal takes off from Massanjore Dam. The canal receives water released by the reservoir of the Dam. Coursing its way through the wasteland is the Mayurakshi canal which holds considerable potentials to bring prosperity to the districts.
4. Parthasarathy has shown that high yielding variety of rice grown in irrigated lands employs nearly one-fifth more mandays of labour per hectare than local varieties usually raised under unirrigated conditions. Ladejinsky also observed that frequent application of water, insecticides and weedicides and double cropping etc. have resulted in more labour demand. See Parthasarthy (1974); Ladejinsky (1969).
5. Of course, some farmers were found to stick to traditional crops. In fact, the irregular and uncertain supply of canal irrigation water was obviously no incentive for farmers to start growing crops which require perennial supply of water. As such, this is not an act of conservatism, rather a rational and calculated response to the inadequate supply of canal water.
6. In the following Table 2.11, t-values (Fisher's 't') for the Farm Business Income between irrigated and unirrigated farms in each size class are found significant throughout. It suggests that in the study area, irrigated farm households on an average receive significantly higher income than the unirrigated farm households. This is true for all classes of farm operators. So, the provision of irrigation facility tends to distribute income in a manner favouring the irrigated farms in general as well as farms in each of the size classes in irrigated areas in comparison to unirrigated

Table 2.11: T-values for Farm Business Income (FBI)

(Rs per acre)

Size class	FBI in irrigated zone	FBI in unirrigated zone	t-value (Fisher's t)
I	II	III	IV
I	2,640.00	652.00	5.42**
II	2,178.00	503.00	5.11**
III	1,343.00	440.00	5.09**
All Farms	1,738.00	372.00	4.38**

**Significant at 1 per cent level .

7. Some detailed studies have already been conducted in understanding the problems of efficiency in the use of irrigation; but only recently distributional aspects have started receiving the attention they deserve from researchers (Bromley et al., 1980, Lenton, 1984, Malhotra et al., 1984 , Abernathy, 1986).

   Inequity problem in irrigation has two dimensions : one, across farm locations , normally head reach versus middle reach versus tail-end locations; another, the farm size-wise distribution of irrigation benefits.

8. It is to be noted that since our analysis is based on data at a point in time, the conclusions should be taken as somewhat tentative subject to further empirical validation with data over time.

THREE

CRITICAL ISSUES FACING INDIAN IRRIGATION

The development of irrigation has been rightly given a key role in Indian agricultural development strategy. Investments in irrigation constitute the single largest component of agricultural investment. Over the past few decades, the state spent a total of Rs. 435 billion on agricultural and rural development including irrigation and flood control ; investment in irrigation alone totalled Rs. 180 billion[1] (Vaidyanathan, 1991). The largest part of state expenditure on irrigation went to construct new and rather large surface systems[2]. The state also gave a substantial amount of assistance to groundwater development. Total gross irrigated area as a proportion of gross cropped area increased from 18.3 per cent in 1960-61 to 33.3 per cent in 1990-91.

But the experience of the past few decades highlights several issues. These are mainly concerning the kind of irrigation works that should be promoted, the relative emphasis that should be given to quantitative expansion, the means of achieving qualitative improvement, the question of equity in the distribution of benefits from irrigation, the recovery of costs from beneficiaries, pricing and subsidy issues, environmental aspects, societal vision of the projects, stabilising role of irrigation input etc.

Choice between Large and Small Works

There is a view that minor irrigation works should be given greater importance than large storage-based canal system which currently absorbs the bulk of public investment. In the choice, the following considerations weighed most. We have a much greater control over minor projects. They are not generally plagued by time and cost overruns. The area covered is limited and so, there should be no time-lag in utilization. The gap between potential created and potential utilised is not essentially a problem. The question of conflict between head reach farmers and tail-end farmers does not arise. In the case of minor irrigation works, the problems of waterlogging and salinity do not exist.[3] On the contrary, the problems relating to heavy investments, long gestation periods, the human, social and environmental issues are very much present in the case of major irrigation projects.

However, it stands to reason that the choice between different categories of irrigation works is not constrained. In any given tract, the effective scope for choosing between large and small projects or between surface water and groundwater is limited by rainfall pattern, topology, geology etc. Again, it is evident that small irrigation tanks are concentrated in the southern part of the Peninsula which received rainfall in the monsoon, but are scarce in tracts depending primarily on the southwest monsoon, even when topography is favourable for local storage. In the latter areas, surplus water from the monsoon months has to be harvested over a wide catchment and impounded on a large scale in order to serve the need during the drier period. Again, in areas commanded by surface irrigation systems, conjunctive use of groundwater and surface water suggests a strong positive complementarity between these two systems. Under these circumstances, it would be appropriate to look upon both the systems as complementary rather than substitute to each other.

New Works Versus Better Use of Existing Works

A second class of issues concerns the balance between creating new works and making better use of the existing capacity of the different projects. It is distressing to note that the potentials created at heavy cost to the public exchequer are not being fully utilised. The government starts far more projects than can be accommodated within the amount of investment available for irrigation. So, projects take normally longer time to complete than anticipated. The average time taken to complete major projects (based on a study of eight projects) is estimated to eleven years to complete, about four and half years more than projected in the original plans.[4] The factors that contribute to the less-than-capacity utilisation can be broadly attributed to the inadequate maintenance and to the poor distribution system. It has been observed that in the absence of the barest minimum maintenance expenditure, the irrigation system is not effective. The poor distributive system, characterised by absence of field channels and other necessary requirements, constitutes a major cause for under-utilisation.

Constraints of financial resources, mutual dissensions among the beneficiaries, litigation over location of field channels, their general apathy, absence of awareness among the farmers were some of the factors that account for virtual non-start of programmes essential for improving the performance of the existing projects. It is sometimes presumed that the under-performance in irrigation utilisation was directly linked to absence of any governmental involvement in respect of improving the system. It emerged that the crucial issue is not as much the growth of irrigation potential as the optimal use of the created potential.

Pricing of Water

The formulation of a sound pricing policy for public irrigation water, particularly in the context of a less developed economy, is beset with a number of conceptual, empirical and administrative problems. It may be noted that the proposition of supplying irrigation water free-of-charge holds little ground. It is true that since irrigation provides effective insurance against droughts, scarcity conditions and even famines and also constitutes an integral component for Green Revolution, it should be subsidized considerably by the states. This seems to be unacceptable, because it has some possible adverse impacts on scarce resource allocation and inter-sectoral and inter-personal income distribution. This free supply of irrigation water involves broader issues of equity. This is because under such a system, the non-beneficiaries belonging to both the rural as well as the urban sectors would be required to reimburse the expenses incurred for provision of the irrigation facility to the beneficiaries.

The argument that since government usually provides a number of infrastructural facilities, it should also, in all fairness, supply the basic infrastructure to the vital sector of the economy does not conform to the principle of public utility pricing. Again, the view that water charges unduly restrict the demand for irrigation which in its turn leads to under-utilisation of created irrigation potential that benefits none but causes enormous social loss, also runs counter to the empirical findings (Ansari, 1986) Again, since benefits of irrigation through increase in crop yields and net farm income are identifiable, irrigation water should be a priced product. Little has rightly suggested that water needs to be sold for a charge, even though the marginal cost of its supply is either zero or near so. Charging for public irrigation water also helps allocate scarce water resources efficiently and minimise the chance of its mis-utilisation. Irrigation tariff can also reduce the adverse income distributional effects that arise out of a skewed pattern of land ownership.

In the context of pricing of public irrigation water, two issues are generally raised. They are : (a) what should be the criteria for fixation of water rates and (b) what should be the method of charging irrigation water i.e. whether water charge should be collected per unit of land irrigated or per unit of water delivered.

In India, water charges on average are very low–amounting to only 2 or 3 per cent of the harvest's value in most states[5]. Against the water rates of 5 per cent of gross income from food crops and 12 per cent of cash crops as recommended by the Irrigation Commission (1972), the actual gross receipts per hectare do not account for even more than 3 per cent of the gross productivity per hectare of irrigated area. Similar problems exist in the case of groundwater irrigation where water rates reflect only about 1/6th of the economic water rates. For example, in Haryana, the water recovery rate for tube-well irrigation is about Rs. 1.25 per unit of electric power consumed against operating costs of Rs. 10 per unit of electric power consumed. With 7 million units of power consumption for all government managed tube-wells, the annual financial loss is Rs. 60 million. Several states (including Andhra Pradesh, Bihar, Gujarat, Orissa, Rajasthan, West Bengal, Madhya Pradesh, Uttar Pradesh etc.) announced revisions during 1981-86; but in some cases, the implementation of the revised rates was held up by the Government. On an average, gross revenue collection per hectare of irrigated area amounts to about Rs. 68 against which the estimated working expenses are about Rs. 216 per hectare, resulting in a financial loss of about Rs.148 per hectare per year. The implication of low irrigation rates with added problems of low recovery is that today the water charges collected do not cover even working expenses, not to speak of depreciation charges and return on investment.

Table 3.1 : Working Expenses, Gross Receipt and Range of Water Rates *(Rs./ ha)*

Name of the States	*Working Expenses*	*Gross Receipts*	*Range of Water Rates*
I	*II*	*III*	*IV*
Andhra Pradesh	1,112	124	99 to 222
Bihar	259	27	30 to 158
Gujarat	4,163	189	25 to 830
Haryana	648	76	17 to 99
Kerala	548	45	37 to 99
Maharashtra	4497	283	65 to 800
Orissa	498	76	14 to 81
Punjab	399	89	20 to 75
Uttar Pradesh	615	64	7 to 230
West Bengal	405	14	74 to 508
All India	877	82	NA

Source : Central Water Commission (Statistics Directorate), Government of India (1991).

Subsidy

In the presence of market imperfection and highly skewed distribution of productive assets, there is an inherent tendency for resources to be misallocated and wealth to accumulate in the hands of a few. In recognition of this, Government of India has combined the traditional role of the State as a provider of infrastructural services with that of an active intervention in redistributing scarce economic resources in favour of the poor. Interventions by the Government may be in the form of subsidy on inputs like irrigation water with the objective of improving agricultural productivity and moderating final prices of foodgrains which are an important item of consumption for the poor.

Some idea about the economic classes to whom irrigation subsidies accrue can be had from the Input Survey (1986-87) and the Agricultural Census data.

It can be mentioned that subsidy in irrigation consists of low water charges levied by states, non-realization of water dues and capital subsidy for digging of wells and boring of tube-wells. Again, the sources of irrigation differ substantially across size of land holdings. It is evident from Table 3.2 that small and marginal farmers rely more on canals and tanks as a source of irrigation compared to large farms; whereas tube-wells have emerged to be one of the important sources of irrigation in all categories of holdings, particularly the medium categories.

Table 3.2 : Irrigation by Size of Holding and Source Net Irrigation Area Source of Irrigation

(Percentage)

	1985-86				
	Marginal	*Small*	*Semi-medium*	*Medium*	*Large*
	I	*II*	*III*	*IV*	*V*
Canals	38.39	37.42	36.25	39.53	39.33
Tanks	9.99	8.31	6.58	4.91	4.17
Wells	10.98	14.84	16.80	19.64	22.51
Tube-wells	31.99	32.09	33.69	34.03	29.21
Others	8.65	7.34	6.67	4.97	4.57
Total	100	100	100	100	100

Source : All India Reports on Agricultural Census (1967-77) and (1985-86), Dept. of Agriculture and Co-operation, Ministry of Agriculture, Government of India.

Table 3.3 : Change in Irrigation by Size of Land Holdings

	Percentage of Gross Irrigated Area to Gross Cropped Area (1975-76)	*Percentage of Gross Irrigated Area to Gross Cropped Area (1985-86)*	*Increase in terms of percentage points*
	I	II	III
Marginal	34.5	39.5	5.0
Small	29.7	34.4	4.7
Semi-medium	26.9	32.2	5.3
Medium	21.0	28.6	7.6
Large	13.4	22.0	8.6
Average	23.7	31.0	7.3

Source : All India Reports on Agriculture Census (1976-77) and (1985-86), Government of India.

It is evident from Table 3.3 that the proportion of gross irrigated area to gross cropped area is the highest for marginal farms followed by small farms in the year 1985-86. Though reduction in some input subsidies in agriculture can perhaps be justified on grounds of correcting distortions in relative prices and thereby introducing greater efficiency in allocation of resources, these increases in input subsidies have also, by and large, led to productivity and income gains on small and marginal farms. Hence, it is sometimes argued that since input subsidies have an elastic consumption and positive yield response, reduction in subsidies in irrigation may not only lead to decrease in consumption of irrigation and hence resultant fall in productivity, but may also have adverse distributional impact (Kumar, 1971).

Effectiveness of Irrigation

What should be a matter of concern is the apparently limited effectiveness of raising the productivity of land. The intensity of irrigated cropping as well as yield of major irrigated crops are much below the levels achieved in other countries and those demonstrated to be feasible under Indian conditions. Moreover, available evidence does not point to significantly faster and sustained improvement of these levels of irrigated land (compared with unirrigated land).

The impact of irrigation on productivity is a function partly of the characteristics of the system and partly of how well water is managed in terms of assurance and timeliness of supplies according to the need of the crops. The marked increase in return to irrigated agriculture that follows the introduction of new cereal varieties and the spread of rural electrification has not resulted in the rapid spread of well irrigation. The problem is how to regulate the number and location of wells in order to prevent over-exploitation of acquifers, to ensure equitable access to water and to keep costs in check. Collective efforts of the users themselves or state intervention to regulate the growth of wells which could, in principle, lower costs and ensure more equitable access are conspicuously absent.

Another important point is that the area under minor surface irrigation system has not expanded much in recent times. The bulk of the expansion of area irrigated by surface water has been accomplished by major and medium works. These works serve anywhere from ten thousand to one million hectares, usually from a single storage through a network of canals maintained and operated by the State. These systems are designed on the basis of available information regarding volume of water flow at the reservoir site, its seasonal pattern, certain assumptions about conveyance and application losses, crop water needs and crop patterns etc. The system designers have to use, more often than not, incomplete and inaccurate information. This may result in inconsistencies and ineffectiveness. The Command Area Development Programmes and modernization of older systems seek to remedy these defects. But many of these defects cannot be remedied only through physical investment; tighter organisation capable of enforcing the necessary discipline is a prerequisite.

It is true that some changes like consolidation of holdings, reshaping of land which affect location, soil composition, boundaries of individual holdings etc. will benefit the irrigators involved collectively by improving the conservation of moisture from rainfall, raising the productivity of water, facilitating the development of wells. But these operations cannot be accomplished on a voluntary basis.[6]

Again, improvement in physical design and control structures only facilitates but does not guarantee better management. Even conjunctive use of water which adds generally to flexibility in water management cannot go far in raising the quality of irrigation without tackling organisational questions.

Equity

Since irrigation is a major source of increased productivity, its distribution between regions and classes is naturally a matter of concern. The extent of irrigation development varies greatly by region. Although, the irrigation ratio increased everywhere during the past few decades, the regional disparities did not narrow. Of course, it may be maintained that disparities in canal irrigation which are mostly in the public sector, have narrowed down substantially. This aggravation of disparities may, however, be due to the development of well irrigation.

Another aspect of equity concerns the impact that irrigation has on the relative income of different categories of farmers. It is well documented[7] that irrigation ratio is inversely correlated to the size of holding. Of course, it is also sometimes argued that this inverse relationship has been weakened because large farmers have superior access to well irrigation and even canal irrigation and also large farmers have the capacity to manipulate the distribution of water in their favour. Of course, the first explanation seems plausible; the second explanation is quite tenuous. There is no obvious reason to assume that all advantageous locations are systematically pre-empted by large land holders. Large farmers are found in all reaches of the canal network.

Another related issue is with regard to system of irrigation used. The more extensive the pattern of irrigation, the more farmers can benefit; because the large is the area that can be benefited from a given amount of water. This appears to be socially preferable. Another line of argument favours intensive use of irrigation. This will increase higher cropping intensity and higher levels of use of inputs and higher productivity of these inputs leads to higher levels of productivity per unit of land. In this connection, it is important to verify whether intensive irrigation necessarily gives a larger total output per unit of water. Clearly, this is relevant, because it is water, not land, which is the binding constraint.

Environmental Issues

Even though irrigation systems in general are beneficial, they are also responsible for certain environmental hazards. These could be broadly listed as follows :

(i) Waterlogging, formation of stagnant water pools and vector breeding places resulting in health hazards;
(ii) Soil degradation, development of salinity and alkalinity resulting in reduction in crop production, uptake of undesirable elements by plants causing long-term health problems;
(iii) Irrigation return flows causing surface water and groundwater contamination and resulting problems of aquatic habitat;
(iv) Drainage channels with aquatic weeds and stagnant water pools; and
(v) Spread of water-borne diseases.

In this connection, it may be mentioned that the loss of forests and displacement of people may cause another environmental problem[8]. The problems of waterlogging and salinity arise primarily from irrational use of water on irrigated tracts. At present, an estimated 13 million hectares are affected (6 million by waterlogging and 7 million by salinity and alkalinity) and the extent is growing. Of course, adequate drainage facilities might help check this tendency. The solutions are recognized but the scale of efforts necessary is believed to be inadequate to the dimensions of the problems.

Destabilizing Aspect

A set of studies, in recent years, have tried to analyse the stabilizing role of irrigation (Mehra, 1981; Dhawan, 1983; Nadkarni and Deshpande, 1982 among others). Mehra, for instance, found that the variability of yield for all the food crops increased in the period 1967/68–1977/78 compared to the early period of 1949/50–1964/65. She reported that in Punjab, the yield variability has been lower, particularly in those areas having a high proportion of tube-well irrigation. In a statewise analysis, she found a strong association between per cent area irrigated, high mean yield and low co-efficient of variation. Dhawan, in a study of Tamil Nadu, finds an increased instability of output in irrigated areas, but largely due to variations in area sown and not necessarily, owing to increased yield variability. He also argued that the stabilising role of irrigation should be studied by comparing irrigated and rainfed farming in the contemporary rather than the pre-HYV and post-HYV strategy of farming. Nadkarni and Deshpande, in a study in Karnataka State, for the period 1955/56–1975/76 observed instability in yields of all major crops, but argue that it is entirely due to climatic and natural factors and that policy instruments are needed to avert increased instability even in areas with assured irrigation.

It may be maintained that much of the output instability in the case of irrigated farming is traceable to area instability; whereas much of the output instability in the case of rainfed farming is traceable to yield instability. How far irrigation in a state is insulating its farm economy against the ravages of drought is a moot question. Emphasis on irrigation in Indian planning emanates as much from its output (and income) augmenting role as from its output (and income) stabilising role.

Societal Vision

Some states in India are very often reeling under a severe scarcity of water for drinking, domestic, industrial and agricultural use. It is sometimes maintained that the root cause of the recurrent water crisis is the failure to grasp the nature of precipitation and absence of sound methods and medium of storing water for various uses. Quintessentially, without understanding the interconnection and interdependence of various ecological entities like land, water, flora, fauna and human along with human syndrome, no worthwhile water planning is possible. But unfortunately, this is the most conspicuously neglected aspect in the prevailing water use projects planning and policy in India.

In fact, the over-domination of the engineering-centred capital-intensive project-oriented approach in water planning is the bane of the crisis. The societal vision is totally missing and the technical and technological solutions are perpetrated, perhaps owing to the lobby of engineers, contractors and their political godfathers (Desarade, 1995).

This is but a bare outline of the major issues that affect India's irrigation system. Given the extraordinary diversity of conditions and complexity of the problems in India, solutions have to be derived from a very careful and detailed analysis of specific situations.

NOTES

1. The figure is based on computations by the Centre for Monitoring Indian Economy, 1988, Vol.1, Table 7-12-1.
2. On an average, nearly one-fourth of the public investment in irrigation during 1950-84 went in to major and medium works.
3. The bias with regard to (over) estimation of area irrigated is more pronounced in the case of wells. The potential of well irrigation is estimated by the number of wells dug and the pumpsets installed. This method, of course, does not consider that a significant proportion of wells is located in the command of surface systems and hence, does not add to irrigated area. Again, a well may need a smaller direct investment per unit of area than surface irrigation. But it requires supporting investment for extending power lines and expanding capacity to generate power. See Deepak Lal (1972). 'Wells and Welfare : An exploratory cost-benefits study of the economics of small scale irrigation in Maharashtra', Organization for Economic Co-operation and Development, Paris, for a brilliant and detailed explanation.
4. Political leaders have a strong tendency in promoting new irrigation projects. It is observed that public irrigation projects are a means of gathering and consolidating political support. In a system based on periodic elections, this is a particularly powerful motivation. Ongoing projects, sanctioned by another minister, give less political mileage than new projects.
5. The Government has spent more than Rs. 600 billion at 1988-89 prices on canal networks during the last forty years. Today, the direct recovery from farmers towards the cost of canal waters is only a small fraction of operational and maintenance expenses, not to speak of capital costs. The low-cost recovery is starving State exchequers and irrigation Department. As a result, the minimum

necessary repairs remain neglected. In 1972, the Irrigation Commission recommended that the price of canal waters should account for 5 per cent of the gross revenue of farmers in the case of foodgrains; while for cash crops, it should be near 12 per cent. The present reality, however, is that water charges probably average only around 1 per cent of the gross revenue of irrigated farms.

The Vaidyanathan Committee Report, Report on the Pricing of Irrigation water, GOI, 1992 gives the following State average percentages of water charges to the gross revenue of irrigated farms:

West Bengal	0.1
Tamil Nadu	0.1
Andhra Pradesh	0.4
Karnataka	0.8
Punjab	0.9
Bihar	1.1
Haryana	1.6
Orissa	1.7
Maharashtra	1.9
Gujarat	2.2
Madhya Pradesh	2.6
Rajasthan	2.7
Uttar Pradesh	2.9

6. The importance of the problem was maintained strongly by Minhas (1970) in his article, 'Rural Poverty, Land Distribution and Development Strategy' in *Indian Economic Review*, 5(1), April, pp. 97-168.
7. This inverse relationship is seen in the National Sample Survey estimates of irrigation ratio by size of holdings in its landholdings survey, but the relationship may be weakening over time. The relevant data are set out by Vaidyanathan in 'Irrigation and Agriculture Growth', in *Indian Journal of Agricultural Economics*, October-December, 1987, pp. 503-527.

FOUR

CONJUNCTIVE USE OF SURFACE AND GROUNDWATER IRRIGATION

Some Issues

The limited success of new biochemical technology of crop farming in India has been linked *inter alia* with farmers' access to dependable sources of irrigation, notably individually – owned tube-wells and pumpsets. True, these are ideally suited for meeting the exacting water requirements of higher yield variety – wheat and paddy. The lobbyists of well irrigation argue by focussing the sheer merits of well irrigation as well as in a negative manner focussing on the demerits of major canal irrigation works : the problems of waterlogging and salinity in canal commands, dislocation of population from lands submerged by reservoirs, high siltation rate of reservoirs and loss of forest cover during and after the dam building stage etc. Again, as between minor and major irrigation works, the minor irrigation lobby has been ascendant, because the capital requirements of minor irrigation are one-fourth of the corresponding requirements for major irrigation works. On the benefit side, it is claimed that groundwater -based farming is twice as productive as surface irrigation-based farming. The impression has been created that the cost-benefit dice is loaded in favour of groundwater irrigation. This impression has acquired credibility. But in fact, the Indian conditions, major versus minor or surface versus canal is not a substantial issue. Both groundwater and surface irrigation are interlinked hydrologically, giving rise to powerful external effects. The role of canal seepage in the economy of wells established in the region of low rainfall where natural availability of groundwater is poor, has been widely appreciated. Dhawan examined the existence of a high rate of return from wells benefiting from canal seepage compared with those not benefiting from it.[1] In the presence of such externalities, it would not be prudent to talk of getting advantage of comparative costs and return in planning. Again, being preoccupied exclusively with the development, either source of irrigation can ultimately be costly. All our irrigation capabilities need to be developed, subject, of course, to technical and economic conditions, because even after tapping all exploitable water resources, a large part of cultivation may have to be carried out without the aid of irrigation.

Again, today low productivity of canal irrigation is well below the corresponding levels under wells. With regard to adequacy of irrigation, a protective canal by the logic of its conception cannot usher in an era of high crop yields. Even a canal designed for productive farming cannot deliver the goods. The actual delivery of water at the final outlet is below the designed value due to say, successive transit losses. The beneficial role of the water losses of the canal seepage has not received adequate attention. In fact, canal water improves the groundwater-based agriculture.

It may be relevant here to have an idea about the impact and the problems of conjunctive use[2] of surface and groundwater faced by the farmers. It has been observed that farmers have invested in wells to grow the crops of their choice. If we look at Table 4.1, we may find some differences in crop pattern between well and non-well owners per se. The impact of wells or other sources other than canal is more pronounced in terms of area under perennial and heavily irrigated crops like paddy, wheat, pulses, oilseeds and sugarcane.

Table 4.1 : Cropping Pattern and Yield

Crops	*Well Owners*		*Non-Well Owners*	
	Area (in % to gross sown area)	*Yield (in qts)*	*Area (in % to gross sown area)*	*Yield (in qts)*
Paddy	44.8	19.8	31.5	31.6
Wheat	8.5	10.5	1.5	6.8
Oilseeds	14.5	8.0	7.5	7.0
Pulses	10.0	12.8	5.5	8.5
Potato	08.0	10.0	3.5	8.5
Sugarcane	3.5	54.5	1.0	30.8
Vegetables	8.7	3.5	9.5	3.0
Others (Coarse Crops)	3.0	4.8	40.0	3.5

It is evident from the above Table that paddy is the most important crop of the Mayurakshi Canal Irrigation Command area, accounting for a predominant proportion of the whole cropped area. The other crops grown in the area are wheat, oilseeds, pulses, potato, vegetables, sugarcane etc. It is interesting to observe that while the area under Kharif paddy has remained unchanged in all the blocks, the area under boro paddy shows a remarkable increase in the blocks like Sainthia where installation of tube-wells for alternative source of irrigation water is made. The other crops which recorded a marked allocation in acreage are wheat, sugarcane, mustard, seed etc. The Table 4.1 also reveals that the acreage under all the provisional crops has been higher in the areas where there has been provision of well irrigation . In areas where canal irrigation is the only source of irrigation, acreage under different crops does not demonstrate any change in acreage allocation. It also emerged that boro paddy and rape and mustard show not only a remarkable increase but also a continuous rising trend.

It is again clear from the Table that yield rates of different crops for the farmers following conjunctive use of canal and ground water are higher than those who use wells for irrigation. In fact, the yield differentials obtained by well owners do not look significant enough particularly for some crops like boro paddy, sugarcane, vegetables, potato etc. Surprisingly, it also emerged in the course of investigation that there were also some non-well owners desirous of growing some water-intensive crops who sometimes enter into an agreement with the neighbouring well-owners for supplying water on some cash payments where canal water is not available. It is also observed that some non-well owners have obtained higher yields. This was predominantly because of the fact that sometimes some farmers grow certain crops on experimental basis in small plots; yield rates are likely to be higher because of better care and supervision and effective management practices.

So, it follows that there are two major reasons for conjunctive use of water. In the first place, normally, the farmers in the command area are not expected to visualize the need for a supplementary source of irrigation, because they expect that canals would provide adequate water for the crops they intend to grow. But, in reality, as we see, water supplied from canal is generally found to be inadequate ultimately and thus not dependable. All these necessitated the development of well irrigation as a supplementary source of canal water. Secondly, the factors like the desire to shift the cropping pattern from traditional crops to modern high-yielding hybrid varieties on the one hand and the preference to grow commercial crops like sugarcane on the other hand have induced farmers in the command area to adopt well irrigation.

The well owners had been asked through an open-ended questionnaire to indicate the factors which, according to them, were responsible for acquiring well when canal water is available. The factors were ranked according to their degree of importance. The factors ranked 1, 2, 3, 4, 5 etc. were given weights 5, 4, 3, 2, 1 etc. respectively. The total rank score for each factor was obtained by multiplying the frequency the factor was ranked first, second... with the respective weightage and adding them up. The reasons were then arranged in ascending order of importance on the basis of total rank score and finally ranked. The total rank scores have been given in parentheses against each item in Table 4.2. That canal

Table 4.2 : Factors Responsible for Well Acquisition

	Factors	*Rank (Score)*
(i)	Canal water supply is not timely, dependable and adequate	1(651)
(ii)	To grow water-intensive crops	2(584)
(iii)	Location of land in the tail-end	3(410)
(iv)	To avoid the problem of waterlogging	4(285)
(v)	It is economic	5(210)

water is not available adequately in a dependable manner on time is recognized as the most important factor that motivates farmers to acquire well. Other factors of importance duly perceived by the farmers are generally the desire to grow water-intensive commercial crops and the undesirable location of the plots of land. Economy and environmental consideration are not very significant.

Well-owing farmers were again asked to give an account of the problems faced while constructing wells. The results are presented in Table 4.3.

Table 4.3 : Constraints for Ownership of Wells

(Figures in percentage)
(Farm-size-class hectare)

Problems	*Upto 2.0*	*2.01–4.00*	*4.01–6.00*	*6.01 & above*
Brackish ground water acquifers	100	98	95	100
Lack of capital	80	76	71	60
Uneconomic and scattered land holding	100	90	80	70
Lack of Power	100	100	90	85
Lack of technical advice	100	85	80	78

The above Table revals that institutional finance and power supply are the two major problems faced by the farmers, closely followed by lack of technical advice on groundwater acquifers. A negative relation between farm size and the reporting of inadequate finance complicates the problem. Small farmers are more hard pressed for money to invest in well. (The farmers in the study region are found to depend on institutional support for investment.) The next important problem is the supply of electric power or price of diesel/petrol necessary for lifting water. Sometimes, some electric pumpset owners complain about erratic power supply, theft of equipment and parts of the pumpset. Some farmers also faced the problem of hard-rock bottoms. This problem adds to the cost of well construction. Another important consideration is the fact that all the small farmers have recognized uneconomic holdings as one of the important reasons for not acquiring wells.

A careful look at the Table reveals that some of the hurdles faced by the farmers like hard-rock bottoms are natural and external and beyond their capacity to overcome. But some other problems like lack of capital, non-availability of technical advice, power shortage, size of land holdings might be overcome. Since uneconomic holding is an effective deterrent for possessing wells, consolidation of holdings may be looked upon as a panacea for this problem. One may think of community wells to extend the facilities to small and marginal farmers where lands are in fragments.

But consolidation of holdings appears to be a sensitive issue in the context of the existing socio-economic scenario of the country. Big farmers are not in favour of consolidation of holdings, because they want to have their control at different places. Big farmers continue to keep hold on water distribution and other related social dynamics if the plots are scattered. On the other hand, the small farmers' apprehension is that influential and big farmers may take good quality of land, leaving only infertile lands at the tail-end areas for them.

Again, it was observed that farmers in the study area did not strictly observe the stipulated norms with reference to the distance to be maintained between canals and proposed wells laid down from time to time by the canal authority. This led to proliferation of wells in some areas. Some farmers have complained about inadequacy of water in the wells during canal closure period due to the high intensity of wells in certain pockets.

It is true that benefits from conjunctive use of surface and ground water are significant. But their extent depends to a great extent on scientific planning and proper integration of canal operational procedures suitable to soil conditions in the command area. It has been observed that near the head and along the mains and branches, the use of canal water tends to be very intensive and groundwater irrigation is resorted to only sparingly. Field visits to some villages in some blocks in the district of Birbhum revealed substantial disinvestment in wells in the head reach areas. Low water rates for canal irrigation, abundance and ease of canal water supplies and facilities of deferred or sometimes non-payment of canal water tax were among the oft-cited reasons for the declining preference for groundwater irrigation in the head reaches.

In order to establish a long-term water balance equilibrium over the entire Command, it would be suggested that the canal irrigation user near the head and along the main and branch canals should be discouraged. On the contrary, groundwater use is required to be encouraged. Because this would be an antidote to water-table buildup and so, the most effective way to secure an extensive spread of canal water resources. Achievement of this at the head and along the main branch canals automatically improves the services to the tail-enders as much as a canal system (with a certain configuration).[3] And in the process, it ensures as much equity as possible in the spatial distribution of or access to water.

Some economists like Professor Dhawan who argue in favour of designing the canal systems for extensive rather than intensive irrigation make great sense. But to operationalize this suggestion, it is not the designing only which is the whole answer. If it is designed nicely, some villages in MIC system cannot supply water to a large number of peripheral villages where the minors and the sub-minors have hardly ever become wet. In this context, elaborate drainage of the works and lining of canals are widely recommended as a precondition for optimum use of water and land resources in a canal command. Drainage removes unwanted water to areas where it cannot damage land productivity. Lining of canals reduces seepage and makes more canal water available to far-flung areas where water may have high value. As such, in the process, it obviates the need to make huge investments in drainage and lining, minimizing thereby the diseconomies of waterlogging and soil salinity.

It is true that the existing canal system management policies and practices militate against the water balance equilibrium necessary to use water and land resources efficiently in a canal command. There is a significant lack of co-ordination between managing the canal system and those in charge of groundwater development. Private investments in wells are generally governed by the relative availability, reliability and economies of groundwater compared with canal irrigation. When the per crop-hectare charges for canal water are substantially lower than cost of groundwater irrigation, they naturally encourage profligate use of canal water, choice of water-intensive crops and strong preference for canal water irrigation in areas close to the head and along main and branch canals, where it is reliable and plentiful. To encourage groundwater irrigation is largely a question of changing the relative economies. It is also observed that no special concessions or subsidies are offered either on capital or operating cost to potential well-owners in the head reaches and along the main and branch canals. Again, as a means to promote conjunctive water use and to act as an antidote to water-table buildup, tube-wells should have to be concentrated in areas suffering from already high or rapidly rising water-tables. A major area will be to alter the economies of canal versus groundwater use specially in these areas. A substantial increase in canal water charges and their effective recovery would help make groundwater irrigation somewhat less unattractive to farmers who can choose between the two. There is a need to regulate the groundwater development activity in the rural areas. The leeway available to do this is limited in the absence of volumetric pricing of canal water which is assumed to be beyond the realm of practicability at the moment. Again groundwater development should have to be encouraged by removing all restrictions and minimising the hassles involved in establishing wells.

India is truly approaching a crossroads when it has to choose a new path for achieving sustainable growth in irrigated agriculture. This path leads towards improving the operational performance of existing surface system by achieving conjunctive use of surface water and groundwater. This is largely possible by improving the use of existing groundwater pumping capacity. The externalities provided by canal irrigation projects are widely recognized. Awareness of the diseconomies imposed by waterlogging and the consequent buildup of salinity on top soil has contributed in substantial measures to the growing disenchantment with major irrigation projects. External economies of canal irrigation are also recognised. However, it is sometimes rightly argued that the benefits that may justify further investment in canal projects originate from the highly satisfactory effect of these projects on the productivity of private investment in groundwater irrigation.

NOTES

1. By and large, economists in India have not appreciated the two-way linkage between groundwater and surface irrigation. Consequently, the output-augmenting role of investment in surface irrigation works in general and major canal irrigation in particular has been only partially understood. Although economists have identified private wells, specially tube-wells as a major force in the successful propagation of HYV seeds, they have largely failed to highlight the role that canal seepage plays in sustaining groundwater-based farming in regions of low and medium rainfall. B. D. Dhawan and K.J.S. Satya Sai in their article 'Economic

Linkages among Irrigation Sources: A Study of the Beneficial Role of Canal Seepage' have examined the economic benefit from canal seepage of surface and groundwater interaction in low and medium rainfall regions with the help of empirical data from two cases: one in the Punjab and the other in the Mula Project in Maharashtra.

2. 'Conjunctive Use' of water is a system in which one system complements and supplements the other to compensate for the inadequacies of both the canal water and groundwaters. It is a development and management of multiple water resources in a co-ordinated manner so that the total yield of a system over a period of years exceeds the sum of the yields of the individual components of the system resulting from an unco-ordinated operation.

3. As groundwater irrigation increases in the head reach and along canals, more canal water will reach the peripheral areas. This would achieve more extensive spread of canal water and also improve recharge to the groundwater reservoirs to the peripheries.

FIVE

MANAGING CANAL IRRIGATION

At the outset, we propose to provide an overview of some of the key issues in irrigation development and water management. Some key properties of irrigation water may be stated as follows.

Irrigation water is an input used for agricultural production. However, unlike some other conventional inputs like fertilizer, pesticides or seeds, it has several specific characteristics which differentiate it (Hirashima and Gooneratne, 1990).

1. Irrigation water is transformed from a free good to an economic good through capital stock in the form of dams, canals, reservoirs, tube-wells etc., although there is no change in either its quality or morphological characteristics. Irrigation water made available through a public irrigation system is perceived as a public good; while underground water, for instance, pumped up by private tube-wells is regarded as a private good.
2. Irrigation water is a divisible input for its users and, like seed and fertilizer, is neutral to scale. However, because of its relationship to the agent of construction, nature of capital used for its construction as well as the relationship between the owner of such capital stock and user of irrigation water, it has a scale bias comparable to that of farm machinery. Moreover, the supply price of irrigation water differs substantially depending upon its location, the type of structure and the technology used, which is not the case with the other inputs mentioned.
3. Irrigation water is a 'leading input' for technological progress in agriculture.
4. Irrigation water has been shown to increase the productive capacity of land as well as its value by improving the 'permanent nature of land'.
5. Irrigation water is not simply an input of agricultural production. It bears a strong relationship to social organisation.

Despite the remarkable expansion of irrigation in India during the last several decades, there is extensive underutilization in canal irrigation.

This problem constitutes a serious lacuna in irrigation planning so much so that this issue often makes one feel that the major problem is not as much of the growth of irrigation potential as the optimal use of the created potential.

Analysing the factors that contribute to such under-performance, we recognize, in general, two major deficiencies being particularly important in West Bengal. This could be attributed largely to (a) the inadequate maintenance, and (b) The poor distribution system (Satpathy, 1984).

Inadequate Maintenance

It has been observed that in the absence of regular repairs to the unlined canals and distributories, there occurs considerable loss of valuable water.

Poor distribution system

Apart from the deficiencies in the maintenance of the irrigation works, poor distributive system, characterised by absence of field channels constitutes a major cause of underutilization. Some economists like Easter (Easter, 1977) have observed that the failure of the farmers to construct field channels within the irrigation commands is the major reason for under-utilization of water from irrigation projects. In the absence of field channels, the prevalent practice of irrigation is to flood the upper lands and allow water to flow down to the fields below by gravity. Under this field-to-field method, water floods over the whole of the outlet passing directly from one field to the next instead of via water courses along field boundaries. Under this system, the irrigation authorities are much concerned with acquisition, transport and storage of water but less with its distribution. Its responsibilities end at the point, where water is supposed to leave the outlet and enter the farmer's water course. The farmers in their turn are expected to excavate field channels and drains and make arrangements on their own to share water and carry out necessary repair. But because of inefficient water utilization practice, the investments on irrigation do not yield commensurate benefits. Certain problems of the gravity flow system may be stated as follows.

In the first place, the gravity flow system leads to colossal wastage of water. The pernicious flooding practice causes less than capacity utilization with actual area irrigated falling short of the capacity of the project.

Secondly, under this system, the quantity of water delivered to any field becomes a function of the distance to the field from the water outlet. This naturally raises the issue 'in the distribution of water as to who gets what, when and where'. This is because under the gravity flow system, the privileged top-enders satisfy their needs first before allowing water to their less fortunate neighbour below. Head-enders may receive adequate and even more than adequate amount of water. As such, the tail-enders often receive less water, less reliably and in a less timely fashion that those near the top.

Thirdly, the farmers do not want water flowing through their fields, particularly while fertilizing, since they are afraid that the fertilizer would be washed away. This often leads to closure of the outlet that denies water to the fields below.

Fourthly, continuous flooding of water in upper land leads to excessive seepage and waterlogging in the lower lands in the absence of drainage. Thus , the very lands which are sought to be benefited by irrigation lose their long-term productivity. At the same time, the low lands run the risk of waterlogging and ultimate salt effervescence. On account of waterlogging, some quite fertile tracts of low lands might go out of cultivation.

Fifthly, the absence of field channels often stands in the way of introduction of an optimum crop pattern.

India has been undergoing deep societal and socio-economic transformation, characterized by fast economic growth, specially in the industrial and service sector, liberal macro-economic policies, development of trade, reforms and privatization in the public sector and institutions, development of civil society and growing awareness of environmental issues and problems. These profound changes in the environment call for a deep transformation of the irrigation subsector, most notably an improvement of its economic, technical and environmental performance, diversification of product and cropping patterns, changes in management systems and its structures, financial and fiscal sustainability (Facon, 1997).

Four main issues are more directly related to the irrigation subsector. These are:

(i) Population growth and national food security

(ii) Water scarcity

(iii) Improving the performance of irrigation schemes

(iv) Degradation of existing infrastructure.

The adoption of rehabilitation/modernization policies is in response to these issues. These issues are, of course, interrelated by causal or hierarchical relationships. Put together, these four issues provide a general framework which is stated in detail in Table 5.1.

Table 5.1 : Main Issues in Irrigation Scheme Modernization

Issue	*Response*	*Measures*
Population growth and national food security	*Increased and secure food production*	*Augmenting irrigation land and/or cropping intensity*
Water resources constraint	Potection of water resources, water resources management, develop conjunctive use, improve water use efficiency, develop appropriate cropping pattern	
Irrigation system performance inefficiency	Optimise water use match supply with and conveyance issues water management	Rehabilitation of infrastructure improving flexibility
Degradation of Irrigation infrastructure	Rehabilitation/ modernization programme, improving financial base, reducing operation and maintenance costs	Improving income from farming, increased water charges, improving design, increasing accountability of agency

The modernization/rehabilitation of irrigation schemes as a part of a broader transformation of the water and agricultural sector responds to a complex set of institutional, technical, operational and economic issues and would consist of a complex set of institutional, technical, operational and agricultural changes, generally associated with changes in water pricing and cost recovery.

Major irrigation is an essentially post-Independence phenomenon. But the potentials under medium and major projects have not been fully utilized. By 1985, about 83 per cent of the potential created has been utilized (Government of India , 1990). This gap between potential created and utilized has been a matter of concern for the planners. Even in the case of potential utilized, we are, in fact, very much in the dark as to the nature and distribution of effects the irrigation has upon the peasant economy in general. It is also frequently observed that the investment of scarce capital resources in irrigation projects results in low return. This has already led to the emergence of an anti-major irrigation lobby in the country. The problems arise out of unduly longer periods of construction. Delay in completion of the project results in cost escalation. There are projects which are under construction for more than 30 years. Naturally, one raises the question of cost-effective benefit from major irrigation. Again, inter-state water disputes contribute to the spread of available resources, thinly spread on multiple projects. Sometimes, the state Governments tend to start several projects simultaneously. This, more often than not, has led to enormous delay in the construction of projects and consequently the cost-ineffective returns from the investment. Again, political leaders have a strong interest in promoting new irrigation projects for different reasons. Public irrigation projects are a means of gathering and consolidating political support. In a system based on periodic elections, there is particularly powerful motivation. Ongoing projects, those sanctioned by another Minister/Administration, may give generally less political mileage than new projects. This militates against objective, comprehensive and long-term planning of irrigation (Vaidyanathan, 1995).

The need for irrigation management has never been more important than it is today. As it has been clearly noted, the expansion of irrigation is essential not only to increase agricultural production to meet the food and fibre requirements of fast increase in population, but also to ensure balanced regional development. In spite of the fact that irrigation potential has increased, the productivity of irrigated crops seems to have been much less than what it should have been. As Bottral observed, there is much evidence to suggest that an increase of as much as 20 per cent in total production can be achieved merely by improving the performance of irrigation systems through 'management reform' (Bottral, 1981). For instance, the average productivity of irrigated crops is 1.6 tonnes per hectare as against 5 or 6 tonnes expected per hectare. Again, maldistribution of water in major irrigation projects is believed to have affected adversely the environment – biological as well as social system. Water-related conflicts and litigations seem to have increased in irrigated areas.

Till recently, irrigation has been considered synonymous with the construction of dams and canals network. The distribution and management of irrigation water has received much less attention. It is said that the neglect of this vital aspect of irrigation has led to several adverse effects. Some others are of the view that many irrigation projects have been designed and managed exclusively from an engineering point of view (Thavaraj, 1982). The socio-economic and environmental dimensions do not seem to have been integrated with the technical designs. So, it would be appropriate to characterize irrigation projects not merely as a hydraulic system to be run according to engineering principles, but as socio-economic system where all participants –farmers, managers and politicians– attempt to maximize their private interest to the cost of social good (Carruthers, 1987).

Irrigation is a complex social process. Caste, class and factional ties and antagonisms become relevant in the delivery of goods or services to the community.

Irrigation systems tend to be viewed as complete in all respects where the channels reach the fields. As such, one should not underestimate the extent and nature of irrigation water management problems. Irrigation technology should be looked upon as a set of human activities which aim to augment and control the supply of water to the soil for the purpose of enhancing the production of crops. So, the planners, the operators of irrigation system and the farmers feel the need for management of water to minimize adverse effects and increase positive gain. Irrigation management as such implies a process in which individuals or institutions are supposed to set specific broad objectives for the irrigation system. It is essential to identify and mobilize and appropriately utilize resources and to set up right conditions in order to achieve these objectives, of course, without causing adverse effects (Lenton, 1988).

It should be essential to note that technical feasibility is the overriding factor in major irrigation planning. But at the same time, political and other considerations, particularly in a democratic polity, converge to bolster technical feasibility, economic viability and social acceptability of the project. Different social, economic and environmental issues may crop up leading ultimately to non-realization of objectives. There are different aspects like water distribution methods, cropping pattern, water management practices and farmers' participation in the overall irrigation management which determine the efficiency of a project. Precisely, for the successful management of an irrigation project, it is imperative to intertwine both the technical and social processes. While executing the apparently well-conceived plans to accomplish the goals and objectives set in the Project Reports, certain important points merit consideration.

One of the important dimensions of irrigation projects that influences management strategies to a significant extent is its protective nature. Protective irrigation is defined as the form of irrigation where canal irrigation is designed to be used to supplement rainfall in meeting crop-water needs. This is done to provide insurance against the total failure of crops by supplying water at critical stages of crop growth – in other words, with a view to protect the rural masses against drought on a more enduring basis; but there are a lot of constraints which the irrigation planners some time seem to have failed to perceive.

It is assumed that the major irrigation systems are designed for continuous flow of water with the assumption that the water will be used during night also. At the same time, to control, monitor and ensure equitable distribution of water, infrastructural facilities like water turn-gates, cross-regulators, measuring devices are essential. But it has been observed that the irrigation structures are not complete enough, particularly at the secondary and tertiary canal system. As such, ensuring efficiency in irrigation management becomes a problem.

Another important issue to be considered in the context of efficient irrigation management is a localisation pattern. Keeping in view the protective nature of irrigation system, the cultural command area of a project is determined on the basis of reservoir capacity and the cropping pattern. Farmers are supposed to know in advance how much water will be available. But in reality, the farmers are not aware of the protective nature of irrigation. Even if area is notified for irrigation, it is generally not made clear to the farmers as to how they should use water and with what frequency. Though it is also assumed that night irrigation will be practised by all farmers, in actual practice, farmers prefer to irrigate during the daytime only. Much less importance seems to have been given to this aspect.

Again, farmers prefer to grow high-yielding hybrid varieties of crops whose water requirements are more and exacting in order to get more income from irrigated crops. And, therefore, the farmers will try to use the available water to their best advantage. It is rather very difficult to prevent farmers from using water the way they want, particularly in an irrigation project which is designed for continuous flow and where water distribution strategies are not well articulated and the physical infrastructure to control and regulate water use is not sufficient. Necessarily, the basic tenet of the protective irrigation policy has been defeated.

Furthermore, the mismatch between theoretically assumed water duty and actual field conditions leads to scarcity of water. In view of scarcity of water (i.e. when water is not sufficient enough to meet crop-water requirements of the specific area), local dynamics are likely to play a major role in water use strategies. And naturally, influential farmers who are also advantageously located at the head reaches of canals normally exploit the situation to their advantage. On the contrary, the tail-enders and the weak farmers may get much less water or no water at all to irrigate even parts of their notified area and hence suffer. In fact, under system of protective irrigation, farmers are generally expected to use canal water very sparingly, only at the critical stages of crop growth or at times of inadequate rainfall and unduly longer periods of dry spells. But it is sometimes maintained that some farmers use more water than what is provided for under protective irrigation and hence water scarcity, particularly for the tail-enders, a characteristic feature of protective irrigation project.

Then, there are some critical questions that need to be taken care of if the protective irrigation policy is to be accomplished successfully. In the absence of appropriate legal and technical restrictions and a scientific operational plan, in the context of widely practised high yielding varieties of technology, in a continuous flow system, under the existing web of economic, social and political conditions, some questions – whether or not modern farm technology to improve agricultural production and the protective irrigation policy can go together, or, would it be possible and feasible to enforce protective irrigation system – are relevant.

The process of an irrigation system operation starts with release of water from the reservoir to the main canal and also to farmers' fields. Different distributories take off from these main and branch canals. Each outlet has several outlets of varying capacity. The discharge of water at the main canal head is based on scientifically worked-out water duty and also certain assumptions on water use practice. But in reality, water is released into canals hoping that it would be distributed properly among farmers. But it has been observed that the accountability of water drawn at various points in the conveyance system, starting from the lowest point, say outlet, to the main canal system, appears to have been much less important. This tends also to create problems in water allocation and distribution. In the command area of Mayurakshi canal irrigation, the accountability at the main system appears to be good; while at the tertiary system, it is almost nil. The accountability of water at tertiary level is equally, if not more, important from the viewpoint of ensuring spatial equity in water supply. Again, under the existing water use practices, it is not generally possible to estimate exactly as to which farmer has used how much water. Individual farmers generally try to use available water to their best advantage with little or no concern for community welfare. A laissez-faire system of water use is found to develop. This also leads to inequitable distribution of water. To achieve equitable irrigation access to all farmers, this general maldistribution provides the challenge for an innovative irrigation management.

It is evident from the foregoing discussion that water distribution policy in canal irrigation system has several operational features which warrant a fresh look for designing an efficient management system. There is a need for a careful fresh look to formulate and incorporate corrective measures to overcome the possible crisis situation when the entire command area gets localized.

Some observations about the performance of Mayurakshi canal irrigation are explained in what follows. The percentage of area allocated for different crop seasons does not seem to have been strictly adhered to.

The ratio of kharif crops in the total area irrigated in each agricultural year is found to be between 26 to 30 per cent as against 40 per cent envisaged in the Project Report. Similarly, rabi crops ratio ranges between 40 to 45 per cent per year as against the stipulated 55 per cent. It shows that rabi crops ratio has increased and that of kharif remained below the stipulated percentage. In fact, it suggests that seasonal crops have been grown in more areas than what is earmarked. This expansion of area under water-intensive crops like boro paddy, sugarcane etc. has consequences on the issue of adequacy of water. In fact, the theoretically assumed localization pattern has not been operationalized satisfactorily.

Hence, the major culprit is the violation of cropping pattern. More than 50 per cent of the total warranted (actual crop water requirement) supply of water is consumed by commercial crops. Area for this crop accounts for about 20 per cent of the total crop area in the agricultural year. That is ,water requirement is skewed towards two seasonal crops. Naturally, the question arises whether in a backward and drought-prone region, it is desirable to use a major proportion of water available for commercial crops. If not, it is essential to identify the kind of design and distribution strategies to be visualized and incorporated in the system to correct the situation.

Water allocation and distribution methods in the command area are not very much different from traditional method of Continuous Water Supply (CWS) in the canals. In some areas, Rotation Water Supply System (RWS) is also practised. The relative merits of CWS and RWS systems may be explained as follows:

> It is sometimes maintained that RWS system is not always useful to farmers. Because, actual supply of water under RWS system is far less that the optimum requirements when compared to CWS system. This is not due to the fact that RWS system provides far less water than is actually required. Some difficulties arise when RWS system is introduced in a canal system designed for CWS. Under RWS system, more water is released in less time. Unless the system is totally rehabilitated, it results in wastage of more water. It may cause breaches of canal bunds and soil erosion.

The RWS system has some advantages. In the first place, it has created an environment when farmers have to become self-disciplined in water use. The frequency of rotation in the outlet is one in a few stipulated days. This will make farmers alert and conscious of water turn schedule. Another advantage is that the RWS system helps in the enforcement of night irrigation. This is because the gap between two irrigations should not be long enough because crops cannot always stand moisture stress for a long period. Therefore, a farmer cannot afford to miss the rotation. Again, controlled supply of water under RWS system has made the farmers realize the need for land levelling and shaping to use available water efficiently. On the other hand, when there is wild flooding of lands, as in the case of CWS system, land levelling is not given due importance. Again, water supply is more stable and dependable in RWS system than the CWS system. Farmers are assured of water on the scheduled rotation day. Again, RWS system encourages the farmers to resort to conjunctive use of canal and groundwater for different regions. The farmers rightly feel that use of tube-wells in canal command areas serves as vertical drains to arrest the rise in water-tables. And, this conjunctive use helps the farmers to grow commercial crops.

On the other hand, in CWS system of irrigation, night irrigation has hardly been practised. And several farmers, after irrigating their fields, divert water to some others' fields due to absence of necessary structure. This adversely affects the crops of other farmers. In the absence of practice of night irrigation, farmers tend to irrigate their crops during day time only. This leads to unnecessary competition and conflicts among farmers. In the absence of well-articulated water distribution policy, local dynamics play an important role in water use pattern.

Let us now make an attempt to reflect upon some of the plausible technical design and managerial interventions to ensure an efficient and sustainable water management strategy.

The study area is designed for protective irrigation to grow only light irrigated crops. But the actual cropping pattern which emerged in the Mayurakshi Canal Irrigation Project (MCIP) area is substantially different from the one envisaged. A large part of the irrigated area is under sugarcane, boro and other water-intensive crop cultivation. As we know, this imbalance in crop pattern has serious implications on water distribution strategies. It is clearly borne out from the results obtained from the study region that imbalance in cropping pattern leads to income inequality among farmers, as farmers at the head reaches grow more of commercial crops and get more income and their counterparts at tail-end continue to grow traditional low value crops and hence, get low income. Farmers may be justified in violating cropping pattern to get higher output, as they are guided by profit maximization principle under a given endowment. But from the viewpoint of productivity of water which has wider relevance to the community, the justification lies in getting maximum output per unit of water used. Preference to water-intensive crops, though profitable to individual farmers, is not socially desirable. When area under water-intensive crops increases in a particular outlet, farmers tend to divert more water than the designed capacity by damaging water control structures and breaching canal bunds.

It is to be noted that the performance of commercial crops in terms of productivity per unit of water is not superior to other foodgrain crops. In the case of pulse and oilseeds, the difference is more perceptible. The violation of cropping pattern by the farmers who are guided by profit maximization principle is disposed more favourably towards commercial crops, negating the very objectives of irrigation projects in drought-prone areas.

So it is imperative to regulate cropping pattern indirectly through control of water supply. Again, it has been observed that most of the farmers do not know the recommended crops. Farmers' rationality in adopting a particular cropping pattern is also guided by their experience and market conditions. So, it is imperative to inform them of the relative merits and demerits of a particular cropping pattern. At the same time, there should be provision for equitable and dependable water supply. In this connection, it should also be noted that there is a need for integrating water discharge and area to be irrigated in the outlet command. This calls for measured discharge of water at the outlet point. It has already been maintained that RWS system would be a better approach to enforce discipline among farmers to use available water judiciously. This would help to increase productivity of water as well as land. So, attitude towards water distribution in the existing project, MCI, needs to be re-oriented through a deliberate public policy of gradual replacement of CWS with RWS. Of course, at the same time, there is a need for rehabilitation of existing distribution system to accomplish RWS in the project in the right way.

It is sometimes maintained that small and marginal farmers are discriminated against in irrigation water use. Influential farmers belonging to dominant castes and classes do use more share in the available water.

The adequacy of water has to be judged with reference to the designed discharge of a given canal or in terms of cropping needs for given cropping pattern. This naturally calls for farmers' awareness and their ability to measure the discharge with the help of the measuring devices, if any, provided. Of course, it may be difficult for the farmers to read it as many of them would be illiterate. So, an attempt is to be made to increase their awareness of some of these issues related to water distribution. Lack of well-articulated guidelines to release water in canals and farmers' ignorance about actual discharges of water make it rather difficult to understand the reasons for inadequate supply of water. None of the farmers has realized the utility of night irrigation. This is to be done. Even if farmers fail to use water during night, the alternative option may, therefore, be to have flexible cropping pattern where water-tolerant crops like paddy may be allowed to grow if soil and other climatic conditions are conducive. Another alternative is to encourage farm-ponds at appropriate places in order to store water during night to use the same during daytime.

A majority of farmers do not have any idea as to how a farm-pond will be useful to them. Attempt may be made to construct farm-ponds on experimental basis at different points in the canal command area to demonstrate their usefulness. Of course, in the case of small and marginal farmers and if the plots are fragmented, the farmers may not afford to have farm-ponds of their own.

Sub-division and fragmentation of holdings in the command area is believed to be one of the main reasons for unsatisfactory implementation of water management programmes, including conjunctive use of water. Consolidation of holdings, of course, is a very sensitive issue in the context of the existing socio-economic set-up in the countryside. The big farmers are not in favour of consolidation, because they are interested in having their control at different places. So, they believe that they will continue to keep their hold in different areas and water distribution if the plots are scattered. The small farmers are not also interested in the consolidation programme; again, because due to this, big farmers may take good quality of lands; the small and marginal farmers may be left with infertile lands at the tail-end areas. Nevertheless, efforts should be made to consolidate the fragmented lands. The small farmers should be assured of their plots to be located preferably in the head and middle portions. The ultimate success depends to a great extent on the political will to implement it.

It has been observed that if any irrigation system does not give adequate attention to the distribution of water along primary, secondary or tertiary channels or field irrigation channels, this may lead to mistrust and misunderstanding between farmers and irrigation officials. The irrigation officials are found to maintain that water duty plays a crucial role in accomplishing an efficient water distribution system. In the absence of any scientific field study, it is difficult for the officials to judge the correctness of water duty. Some officials feel that separate estimates for water duty for kharif and rabi seasons should be made. The violation of cropping pattern and the consequent increase of area under water-intensive crops are found to be responsible for non-realization of estimated duty. The officials also find it difficult to enforce cropping pattern. Farmers hardly follow rotation of crops to maintain soil health. They simply cultivate the same crops on the same plot of lands for years. To take more water with a view to irrigate the land in less time, farmers think that the structures obstruct the flow of water for tail-enders and hence they feel tempted to irrigate only during daytime for watering water-intensive crops. Structural changes are to be made. It is believed by some that demand-based irrigation water distribution system should be introduced. This will help the irrigation officials to plan for water budgeting and allocation. If the present system continues, proper discipline in water use cannot be ensured. Again, improper use of water leads of waterlogging and other adverse effects of irrigation. The reasons cited for these are generally like rotation of crops not followed, lack of proper drainage, not practising night irrigation, lack of scientific on-farm development, higher intensity of cropping etc. In the light of all these observations, it is suggested that irrigation officials are required to have adequate judicial power for control and regulation of water use, distribution, cropping pattern and other irrigation offences without which it would be really difficult to improve the situation.

Lack of proper co-ordination and understanding between farmers and the irrigation officials seems to be one of the main reasons for less efficient irrigation management. Grassroots executives are expected to maintain a good liaison with the people's organizations. People's organizations are thought of as one of the organizational strategies for ensuring farmers' participation in day-to-day problems of irrigation water use and management. Farmers should not be treated as a substitute for bureaucracy, but as a supplement to it for managing the system effectively.

SIX

PEOPLE'S PARTICIPATION IN IRRIGATION MANAGEMENT

India has a long tradition of farmer-managed irrigation system. Before the advent of the colonial rule in India, about six million hectares were under irrigation, none directly managed by the Government. The ancient Indian system of community-managed irrigation found its way to other countries of the world.

British rule in India, in fact, ushered in a new era of large irrigation projects. This made a remarkable decisive break from management. Large river diversion schemes (like the Godavari-Krishna Delta System, the Sone Canal System etc.) were found not to be administered by the farming communities. In this connection, different Irrigation Acts (like the Bengal Irrigation Act, 1876, the Bombay Irrigation Act, 1879 etc.) were enacted to serve as a guide to the irrigation laws. The reflections of the colonial thinking in the Acts were at variance with the spirit of self-governance. In short, one cannot find any appropriate and adequate provisions for encouraging collectivism among farmers for managing the irrigation systems. On the contrary, the different Acts made a negative contribution to the indigenous participatory process. In sum, irrigation system was administered rather than managed (Singh, 1991).[1]

Necessarily, when India became independent, the country had two opposite systems of irrigation – the farmers managed small irrigation system and the large irrigation schemes totally controlled by the state. The necessity of equality in the use of assets created from public investments like state-owned wells and tanks, large canal systems, roads or goods for public consumption was recognized. But the administrative system of the country did not allow itself to be freed from the colonial legacy. In spite of the Government's proclaimed desire to move in the direction of self-management by farmers, the administrative system failed to take up the challenge. The reasons cited normally for this may be stated as follows. In the first place, the reluctance of the bureaucracy to surrender its power to sanction water to farmers; secondly, the desire not to surrender opportunities for gratuitous income; and thirdly, reluctance to surrender to political manipulations through the medium of local government functionary. In sum, the farmers, in general, in the post-Independent India are found largely to be at the mercy of the irrigation administration.

Irrigation as an activity of controlling and channelling a natural resource (water) operates under stringent physical constraints. The geophysical conditions not only impact on the creation of the irrigation asset in so far as the source of water supply is fixed in nature, but also the conditions of utilization of the asset are also equally influenced by geophysical conditions, like the terrain, the distance between the delivery point and the source, soil conditions, topographic features of the territory under the command of the source etc.

Given thus that there is a strong natural resource base for irrigational activity, given also its critical importance in the process of production, major irrigation works (which also require large financial investment) have been one of the cases of 'public works'. This raises the usual problem of distribution of 'collective' or 'social' goods. Further, while the irrigation assets may be constructed under the public sector, its maintenance and utilization may fall under the purview of private activity and decision-making. Different organizational forms and patterns affecting the utilization and effectiveness of irrigation emerged.

Irrigation is a typical case of activity that generates significant 'externalities'. Some of these are ecological and environmental, especially arising from a major irrigation project. This may bring about changes in the exact ecological and environmental balances, affecting the geophysical characteristics of the region. The creation of major irrigation systems often involves also displacement of population with the consequent political and social distress and tensions. Externalities can also arise, more specifically, intermodally: exploitation of water resources through one mode of irrigation may affect viability or effectiveness of operation of another mode advantageously or adversely (e.g., the introduction of a large canal system may affect ground water resource of tube-well irrigation).

Apart from the technical and geophysical externalities, there are also economic externalities. Introduction of irrigation in a region may induce a number of direct and indirect, primary and secondary economic costs and benefits. Externalities in terms of ecological and technological interactions and the evaluation of costs and benefits ascribable to a project have received increasing attention. However, less prominence has been given to issues arising from the discriminating impact of irrigation on different socio-economic classes of users who do not enjoy equal access to the irrigation sources and are not able, on the one hand, to own private means of irrigation or adopt the more efficient mode of irrigation; on the other hand, they also suffer from the inequitable impact of the adverse 'externalities' as well as from the disabilities owing to poor maintenance and utilization of the irrigation assets.

It is well argued that productivity and effectiveness of irrigation depend upon the provision of adequate and well-regulated supplies of water, matching the needs of particular crops according to the phases of crop growth. Mere institution or creation of an irrigation asset cannot guarantee effective irrigation. The latter does not depend only upon the technological and engineering capabilities of the mode of irrigation. Its operational efficiency depends upon how well-organized is the system of maintenance and delivery. The striking paradox of the co-existence of excess capacity or unutilized irrigation potential in a situation of general irrigation shortage highlights this deficiency. Until recently, more attention is paid to the creation of irrigation capability and very little to the maintenance and delivery system. In a peasantry that is sharply differentiated, the socio-economic hierarchy among the users produces discriminatory consequences in terms of the access to irrigation, its maintenance and utilization. The varying degrees of irrigation use and its effectiveness among different users need to be investigated.

Strictly speaking, in the primitive and anarchic state of affairs, there were exclusively user-managed systems. On the other hand, there is no system which is completely bureaucracy – managed. At some stage, the management tasks are handed over to the users. Practically, these irrigation systems are 'jointly' managed by the users and bureaucracy. In sum, all irrigation works are found to be looked upon as possessing a management situation when the responsibilities are divided between an authority and the users.

Water is appropriated by the users either in groups or individually either (a) as private (individual) property or (b) as common property. When water is utilised by the users as common property, then the systems are normally gravity flows in a large scale canal irrigation system. The systems do not come under purview of social management and generally are appropriated as individual property. But for all practical purposes, water management within the community when it is appropriated as common property needs regulation and improvement and is very much a social policy.

Even though water is appropriated as common property, the users themselves have to construct additional channels for appropriation, allocate the water amongst themselves and maintain the physical works within their jurisdiction. Naturally, the users may have a good understanding amongst themselves so as to carry out all these works systematically.

This may be termed as co-operation for common property management.

But occurrence of this type of co-operative management is not inevitable. In this irrigation system, for example, each individual user expects that others will undertake the construction and maintenance works. At the same time, every user (particularly the rational utility maximizer) may try to appropriate more and more water. Hardin (1968) has shown that common property is inevitably being overused and hence rapidly depleted by the self-seekers. The situation is known as the 'Tragedy of the Commons'. It follows simply that users cannot come to co-operate on their own. Hence, systematic management of common property necessitates intervention by an authority.

Some inadequacies in this approach have been recently identified. A critical review of this approach has been made in a formal way in the form of a 'Prisoner's Dilemma' game. The Prisoner's Dilemna' is a game in which each of two players has two pure strategies – confess (c) and maintain innocence (i). The two players are criminals who have committed a crime together and who are separately grilled by the police. Thus, they are moving simultaneously in a game having one move per player. During interrogation, each of them was promised that he can defect (Sucker) by confessing, for which he will get a financial reward in addition to acquittal. But if both confess (defect), their confessions are not so valuable; instead of a death sentence, they may get life imprisonment. They also know, even if not told explicitly, that if both refuse to indict the other (co-operate), they cannot be convicted for murder due to an absence of witness. What do the prisoners do ? It is interesting to note that each of them, if they choose to act rationally, must confess to indict the other. Whatever, be the strategy chosen by the other, he gets the better deal by using this option. If the other convict does not confess, he gets a financial reward plus acquittal by defecting. If his associate too defects, he still avoids the death sentence by defecting. At the end, both of them find that they have earned life imprisonments by choosing their individual best and in the process have sacrificed a far better option realisable only through co-operation. This 'game' makes it an excellent representation of a situation which brings out that there are situations when everyone may suffer loss even if every individual acts rationally.

But Hardin's formulation of the 'Tragedy of Commons' or the 'Prisoner's Dilemma' has certain limitations.

A long-term consideration often favours the adoption of a co-operative strategy as rational choice by individuals sharing a common resource. The formulation does not take into account the fact that the depletion of the common property resources which promise to benefit the beneficiaries for a long time may be considered a loss by the self-seekers themselves. Hence, on a long-term consideration basis, the farmers consider the adoption of co-operative strategy as a favourable rational choice. They will learn in course of time that mutual co-operation is the best strategy. (This is formulated in iterated 'Prisoner's Dilemma' game.[1])

Again, it is sometimes maintained that it is not always true to assume that in all societies, at all times individuals will take care only of themselves. 'Altruism' may also be a characteristic and may also help to generate co-operation between individuals.

All in all, even self-seekers may desire to co-operate[2]; these users' co-operation is just as well-adopted to modern irrigation as it was for the past.

Concept of People's Participation

Farmer's participation in irrigation can be taken to exist when water users take decisions that determine access to water and its use, according to principles considered worthy by them (Singh, 1991). It cannot be fair to say that farmer's participation exists when farmers are informed of the time water is to be released and their entitlement to it or referral of farmers' problems to irrigation officials and farmers' willingness to co-operate with the irrigation agencies. One is informing and the other is conforming. Both involve passive compliance. In fact, formal participation in irrigation can be said to exist when water users take decisions that influence access to water and its use according to principles considered worthy by them. They are:

- — decisions regarding the allocation and distribution of water
- — procurement, appropriation and guarding of water,
- — development and enforcement of social norms regarding water distribution, settlement of disputes, etc.,
- — responsibility for supervision of a part or whole of the irrigation system,
- — self-renewal of the management group or organization of irrigators created for the above purposes, and
- — two-way communication with the Government agency and between the management group and the beneficiaries.

Most of the above criteria are, to a varying degree, expected to be met in the community irrigation projects.

Why Farmers should Participate

There are some important arguments in favour of people's participation. In the first place, irrigation system, particularly canal irrigation, is a highly inter-dependent system. The ability of one farmer to draw water is greatly affected by the behaviour of those upstream. Naturally, when there is inter-dependence in supply, people are anxious to draw one's share at the earliest. So, insecurity and anxiety become motivators of behaviour. Under the circumstances, even the best managed and well-designed irrigation systems might be abused at the hands of the users. Eventually, the irrigation agency may have to fight a losing battle against poor performance. Against this backdrop, as a measure to help the users feel secure is to involve them in the management of the irrigation system.

Again, some erosion and silting always take place in an irrigation system. Grass invariably grows and hence needs to be removed. Sometimes, irrigation structures get broken. Hence, the system is highly depreciation-prone and as such, in order to keep the system in a steady state, this requires restoration efforts. For any irrigation agency, it may sometime prove very expensive. People's participation in the performance of the irrigation system helps solve the problem more easily and smoothly.

It has been observed that in many projects, the systems, on an average, have much less water to distribute than needed by farmers. Under the circumstances, there is a need for best us of available water. This is largely possible through an agreement with the water users and this can be arrived at through mutual consultation.

Water delivery may not necessarily harmonize with farm needs. Water may be available when it is not needed or water supply may be more or less than water requirement. Beneficiary participation is an essential step in linking the delivery of water with its utilisation. Again, it is not desirable that farmers need destroy irrigation structures and undo field channels. Water rents collected from the areas served by public works do not even cover working expenses. The gap between price and cost has widened. Disputes are required to be resolved within the system by the farmers themselves. This is better than mediating conflicts through bureaucracy.

In sum, there are three kinds of usefulness of farmers' participation (Uphoff, 1986) :

(a) They can share the burden which otherwise lies totally with the development agency ;

(b) They provide a channel through which local knowledge

becomes available to the authorities ; and

(c) Once organized for irrigation tasks, they may accept many other kinds of group works.

Why Farmers are Reluctant

In Mayurakshi canal command area, it has been observed that farmers are sometimes reluctant to assume responsibility for self-management. The reasons for farmers' reluctance or their inability are not far to seek. In the first place, in canal irrigation projects, water is supplied to users by an external agency. Farmers are not involved in the system's operations and management nor have they done anything to create the resource. In fact, they are passive recipients. And at the same time, there is no obligation to conserve the water or to protect the system or even to respect the right of others. In other words, there is no sense of responsibility. There is no involvement or commitment whatsoever.

Secondly, as long as water is available, there is no need for co-operation. If supplies are inadequate and if it can be improved, then water users may attempt to organise collective efforts. Of course, success depends largely upon factors like leadership, community's power structure, local traditions supportive of co-operative action. If there are no pre-existing social bonds which can be activated for collective action, then individuals are left to fight for their own cause. The fact that under some circumstances, the weak and small farmers remain deprived does not agitate other relatively better-off water-users.

Thirdly, in a government-managed canal irrigation system, water is , more often than not, in short supply. So, a system for distributing water among uses is usually required. Such a system usually imposes discipline on all users. This is a fact which is not equally accepted, as all individuals do not wield equal personal power. The irrigation agency seems to be more important (as provider of irrigation water) than the co-beneficiaries. Hence, there is a tendency for development of synergic (mutual beneficial) relationship between some farmers and the irrigation officials. Under these circumstances, co-operation does not necessarily always emerge as a natural response.

Again, it may be explained in the following way. Canal irrigation may be looked upon as a Common Property Resource (CPR)[3] and like many other CPRs, is found a victim of the 'Tragedy of the Commons'. Field channels are found silted and are infested by weeds; their beds and foreshore areas are encroached; their sluices and outlets are choked in the absence of regular cleaning and water distribution channels are either absent or are badly silted and broken. It has been observed that all these problems largely arise from a tendency on the part of the co-users of irrigation water to shirk their responsibility to contribute to the repair and maintenance of the project and its protection and management. In other words, co-users do not co-operate with one another. This non-cooperative behaviour may be attributed to (a) the perception of the co-users that their private cost would exceed the private benefit from co-operation; (b) their feeling that their own contribution is too small to have any tangible effect on the improvement of the performance of the existing irrigation system; and (c) in the absence of good leadership, they do not have any assurance that the other co-users will contribute their share.

The canal irrigation projects with their complex engineering and bureaucratic organizations cannot be successful without the active participation of beneficiary farmers in the management process. Creation of water users' association is considered essential. Farmers' representatives should participate in the development of each season's water allocation plan and monitor irrigation agency's day-to-day performance. In the absence of a proper rapport between the farmers and the irrigation officials, there might be the prevalence of what Hart calls 'the Anarchy Syndrome'.

Reidinger (1974), writing about the Bhakra system in the mid-seventies also suggested the creation of water users' association at the water course level. The associations are to be held responsible for the maintenance of minor networks of the canals. This includes desilting and cleaning of field channels. The association will be of immense help in minimizing the conflicts between the farmers of the head reaches and tail-ends. At times, particularly during scarcity situations, members would be required to volunteer their services for night watch so that the upper reach farmers do not breach water supplies.

Organizations should be first stabilized and then attempt to acquire legitimacy in the eyes of the group members. Acquiring legitimacy as a social institution is an important condition for making the system self-sustaining. For this, two conditions are required to be satisfied.

First, it must be successful in dealing with the primary task of irrigation management in large irrigation projects, i.e., obtaining and distributing water according to the acceptable norms, resolving disputes, enforcing disciplines etc. And secondly, it must acquire social identity as an activity centre dealing with matters of irrigation. Again, acquisition of legitimacy cannot be considered independent of the village power structure and leadership. A divisive village can hardly be expected to support a new activity. Power configurations in a village/society cannot be ignored; for, they contain the forces that affect almost every aspect of community life.

In India, there is a considerable potential for involving beneficiaries in the management of irrigation systems. But very little attention has so far been paid to this task. There is a necessity of changing this orientation. Such a change calls for greater reliance on people. There are certain conditions necessary for co-operation in management of irrigation resources.

In the first place, political will is considered a necessary prerequisite for motivating the people to co-operate. A systematic development can be meaningfully enforced only when such a political will is present.

Secondly, in the choice of technology and administrative policies, only those kinds which match the existing knowledge and skill of beneficiaries are to be adopted. It is true that the common property resource (like canal irrigation) management is not free from technological and social changes. The scope of co-operation among users can be made to vary by changing the technological and administrative contents. The authority may impose a set of programmes which, instead of promoting their participation, may restrict the expression of co-operation. The technocrats may make the farmers the scapegoats and this failure to co-operate lends credence to the myth that the farmers do not like to co-operate.

Thirdly, if users are essentially non-co-operative, a situation of co-operation can never arise. Due to severe caste or kin rivalry, the concerned groups may be committed not to co-operate ever. Sometime individuals move blindly, caring little about the past outcomes; co-operation will then become a chance event. So, the beneficiaries of a common property resource are not committed to non-co-operate between themselves.

Again certainty of recurrent and regular supply of the resource helps to sustain co-operation. At the same time, individual farmer's expectations regarding the availability of water and its productive efficiency should be high.

People's Participation : Issues

People's participation is viewed as a significant effort towards improving the efficiency and effectiveness of development programmes at the micro-level. It is maintained that people's participation may have the way for the people to articulate their felt needs and also to fulfil their needs. Equally important is the active involvement of local institutions which is expected to create conditions necessary for sustenance of the development processes.

The root cause of low rate of realization of objectives as also the skewness in the distribution of benefits of development has been the high degree of bureaucratization and over-centralization, among others. The economic theory provides sufficient justification for decentralization based on the principle of Pareto optimality which aims at maximization of the social welfare. The involvement of the people in terms of resource mobilization and resource use will take cognizance of variations in (quality and quantity) resource endowment, addressing the local problems and facilitating the realization of local aspirations. People's participation in the design and implementation of the operational network and scheme will in the process reduce costs of execution and increase the overall effectiveness of the system.

Decentralization and people's participation are not altogether different terms and the extent of participation itself is a good indicator of decentralization. Advocates of 'decentralization' claim three main virtues for it : enhancement of people's participation, responsiveness and efficiency in governance. In the irrigation system, the term 'decentralization' could mean the management of services and regulatory functions of the bodies constituted for the purpose. In a capsular form, decentralization connotes the transfer or delegation of planning, decision-making and executive authority. To amplify it further, it is democratic in the sense that the source from which power is decentralized has its democratic foundations, i.e., the creation of local units to which power is delegated.

As for people's participation, its need in all development efforts has to be reiterated. In point of fact, people's participation means many things to many people and different things to most of them.

The dichotomous and dialectical nature of participation broken into 'top-down' and 'bottom-up' approaches compounds the confusion further. The former connotes mobilization; the latter concentrates on sharing of responsibilities and benefits by the people for whom the project is built. Thus the participatory approach is often described as 'bottom-up'. It has been nicely mentioned by Norman Uphoff (1992) :

> 'A participatory strategy for promoting sustainable agricultural and rural development process on the assumption that rural people have more to contribute to the development process than just their money or labour power. They have ideas, management skills and organizational capabilities that are needed for development. They are to be regarded as partners more than beneficiaries or (worse) target groups.'

It is also relevant to note that the objectives and modes of participation vary according to the context. For instance, in the subject, a lot of modes of participation have been mentioned—like, participation as contribution, participation as partnership, participation as empowerment, participation as organization. And it is also evident that the complexity of operations as well as risks will increase as one moves from lower mode of participation to a higher one. Again, the sustainability of the impact of the development programmes will also increase as one moves up the ladder.

In the case of 'irrigation management', the process of participation may be viewed both as a 'means' and as an 'end'. People's participation as a 'means' may be construed as a 'social technology' to build up social dignity and social transformation to empower the powerless, as a vital ingredient for the success of rural development to attain other ends; for redistribution of power, a variable of development, as a mystical entity to solve the problems of planning and to secure confidence from public. As an 'end', people should participate in 'decision-making'. According to Miller, participation is much more than these dimensions; it is an article of faith, resource input into development, a human need, a human right and a social process. In the ultimate analysis, it is synonymous with self-help. In sum, as a 'means', people's participation enables the people to fulfil their felt needs and aspirations: and thus, it may be recognized as a 'gap-closing mechanism'. As an 'end', it assumes the form of a 'capacity enhancer' in the sense that people should take part in decision-making, and thereby assume the role of partners in the process of development In point of fact, it may be stated that bodies like water users' body are assigned the role of social mobilizers; and for all practical purposes, these institutions metamorphose themselves into vibrant responsive representative bodies of the lowest stratum.

In the management parlance, the SWOT (Strengths, Weaknesses, Opportunities and Threats) analysis is used as a tool to diagnose the factors internal to the system (the people's participation in management of canal irrigation in the present context) and the external environment having a bearing on the organizations as well as the projects (Sastry, 1990).

Table 6.1 depicts clearly the 'SWOT' of people's participation in canal irrigation management.

It is evident that people's participation strategy has a potential for facilitating generation of employment and development of infrastructure. Our field interactions indicate that conflicts can be resolved by a better understanding. The need of the hour is to strengthen the institutional structure at the grassroots level. The process may dilute the role of public authority in the utilisation of scarce water resources. The co-existence and co-operation of the people's participation with the irrigation agencies at the micro-level are expected to dilute the elite-elite and elite-non-elite conflicts. It may also be inferred that over time, the modes of people's participation are expected to change, i.e., from the role of contributors to that of partners in the development endeavours (Dharam Ghai, 1988).

TABLE 6.1 : SWOT Analysis

Strengths	*Weaknesses*	*Opportunities*	*Threats*
i	*ii*	*iii*	*iv*
Legitimacy for decentralized planning is accorded	Bureaucratic abberations still galore	Interface between users and officials may lead to broader understanding of ground realities	Given the resource crunch, scope for improvement limited
Structural backing is ensured through water-users bodies	Structural weaknesses may get compounded by elite-elite conflicts	Gives a fillip to persons manning local institutions	Chances of strained relations thus the efficacy of structure eroded
An opportunily for dovetailing schemes under the rubric of decentralized planning	The political bureaucracy may hinder the smooth operation of decentralization process	Facilitates preparation of perspective plan/ professionalisation of planning activities at the grassroots levels	People may lose confidence in Govt. capabilities
Emphasis on implementation of works relating to Minimum Needs	Inadequacy of funds restricts the realization of envisaged goal	Enhance organizational abilities/ indentificaion of felt needs provides grater scope for greater autonomy	Decisions biased exclusively in favour of the weaker sections may desist the others sections

For developing an analytical framework for the conception and operation of participatory decentralized framework with people as partners in development, the following aspects merit special consideration.

(a) *Accountability* : The criticism that decentralization leads to deconcentration of power and decentralization of corruption, cannot be stretched beyond a point; since accountability to the public is inversely related to the distance between the decision-making body and the public. The long-term spin-offs of empowering people and allowing people to participate and reposing trust in them may outweigh the economic costs and accelerate the process of social transformation in rural areas.

(b) *Involvement of People* : The nature and intensity of participation need to be articulated and spelt out in concrete terms. In the absence of clear analysis, the terms like collaboration or co-operation may be beset with ambiguity and hence, may not yield tangible results.

(c) *Operationalization of the flow of benefits* : It is to be ensured that the reasonable share of the benefits should accrue to the disadvantaged unit and necessarily, the designers should evolve a monitoring system to ensure the smooth operation of the distributional process.

(d) *Appropriate mode of participation* : Our field experience suggests that potentials exist for a higher mode of participation, i.e., partnership mode which can effectively kindle the interests of the people.

(e) The emphasis should be on evolving a strategy to formulate perspective action plan for which the people and the officials should be integrated. The interface session for training and discussion (more of a learning type) between the people and the officials will go a long way to resolve the local inter-level differences. The technical competence at the micro-level may be found inadequate. The officials may give technical inputs to enable people to derive maximum benefits from the system.

(f) It is well documented that the most vulnerable group in the rural areas is women. During our field visits, we also witnessed the relatively less active involvement of the women. So the solution set should find ways and means of mobilising the priority group and tap the enormous potential of women as prime movers of the delivery system. The women groups may be identified properly and can be associated actively in the system so that they can periodically discuss, identify the problems and solutions and be made responsible to monitor and evaluate the schemes.

One general observation is in order. In the current exercise, based on Mayurakshi canal irrigation experience, an institutional framework is required to be developed in a democratic decentralized context which will not only strengthen the people's participation in the act of planning and implementation of development projects with the increased efficiency but also ensure their sustainability.

If history and experience is any guide, irrigation sources like tanks, canals etc. as a productive resource are product of local initiatives and retain their vitality if their upkeep is left to the people who are proximate to benefit. It is the empowerment of the people in managing the resources.

NOTES

1. Those in control of water supply, particularly in the agrarian societies, did enjoy considerable social and economic power. K. Witfogel (1957) maintained that irrigated farming depended on the effective control over a large quantity of water and this could be realized solely through the organisation and co-ordination of a large section of the community. This necessitated the emergence of a co-ordinating and directing authority which typically emerged into a seat of political power and authority. Hence, the inherent tendency of hydraulic societies to become centralized, despotic states.

2. The success or failure of co-operative action in the past can be expected to have a significant effect on the prospects for co-operative action in the future, by affecting the degree of confidence that individuals have in the likelihood that others will play their part in a co-operative agreement. Success breeds success (and vice versa) in co-operative development. Altruism, social norms and long-term self-interest are mutually reinforcing motives in the appropriate circumstances. See Sargent, M. (1982), *Agricultural Co-operation*, (Aldershot, Gower).

3. Oakerson (1986) identifies three attributes for a common property: namely, jointness in use, excludability, and indivisibility.

   Jointness implies that, within limits, all co-users could derive benefits jointly from the resource, although each user subtracts from the benefits available to the others in the group. Excludability means ability of co-users to exclude potential users unless they pay the stipulated water charges and other fees. Exclusion of free-riding co-users is not ordinarily possible. Indivisibility or divisibility of an irrigation project affects success on failure in its management. Although, theoretically, a canal can be partitioned; but, in practice, it is very difficult to do so. All these features hold good in the canal irrigation projects.

   Apart from the problem of inefficiency accompanying the present predominant mode of resource utilisation in common property resource units, the aggravating ecological crisis is leading to increasing consideration of many natural resources also as common properties. Of late, it is being suggested that people's participation is a desirable alternative. So, not only for efficient resource utilization but also for ecological viability, it is pertinent to evolve strategies necessary to promote and sustain people's participation.

   Again, the common resource management through people's participation should not be based upon the assumption of widespread occurrence of altruism, rather upon the existence of a large number of individuals interested in themselves and then can have a case only if there is a possibility of co-operation to emerge. See Sengupta, 1991, p. 273.

To the extent the Game Theory has been used to counter arguments like that of the 'Tragedy of Commons' and establish the possibility of widespread existence of spontaneously-formed irrigation associations, it is of general interest.

In the area of design, for example, scope for co-operation is much less in flow systems than in irrigation from captive reservoirs, like a tank.

SEVEN

TOWARDS CONCLUSIONS AND POLICY IMPLICATIONS

Public investment in irrigation, deemed imperative for accelerated development in the production of foodgrains, came under the purview of the Five-Year Plans which commenced in 1951. Through the fifties, investment was geared primarily towards the creation of large-scale surface irrigation works. The giant schemes like Bhakra-Nangal, the Damodar Valley, Hirakud etc., were taken in hand soon after Independence. The actual achievement in terms of the area irrigated by the schemes is reported to be around 50 per cent of the target. During the Third Plan (1961-66), emphasis was shifted to implement the earlier schemes which had suffered extensive delay in implementation. During the Fourth and the following plans, investment in new major and medium schemes has again picked up. The government took up in the seventies integrated development of some of the major canal networks in the country as a remedial policy by instituting various Command Area Development Agencies.

The study on 'Managing Canal Irrigation in India' brings out certain interesting trends. Let us now look at some of the specific conclusions that could be derived from the study.

India has an enviable record of engineering accomplishments in the field of irrigation and those accomplishments deserve much of the credit for the nation's impressive growth in foodgrains production since its Independence. The fact that major irrigation is essentially a post-Independence phenomenon is quite clearly recognized from the investment pattern in the successive Five-Year Plans. A lion's share (78 per cent) of such investment has been on major and medium irrigation projects. But potentials created under different schemes have not been fully utilized. There is again another view emerging where it is held that canal irrigation has been considered synonymous with the construction of dams and canal network. The distribution and management of canal irrigation water has received much less attention. Many irrigation projects have been designed and managed almost exclusively from an engineering point of view. The socio-economic and environmental dimensions do not seem to have been integrated with technical design.

Therefore, it is sometimes argued that it is not correct to characterise irrigation project as a purely hydraulic system to be run strictly according to engineering principles. All participants like farmers, irrigation officials and politicians are expected to play their role to maximize their interest to the cost of social goods. But it has been observed that the policies of irrigation water distribution and use appear to have not taken into account responses and understanding of the cultivators. The need for understanding the nexus between the technology per se and its operationalization has a special significance in a democratic system. Again , the delay in completion of major projects normally results in cost escalation. Inter-state water disputes often contribute to the spread of available resources thinly on multiple projects. Most of the irrigation systems in India are found to be administered rather than managed.

Under this problems mix, attempts were made in this study to examine empirically some of the issues and problems of water use and distribution in a major irrigation system, both at the main as well as the tertiary levels.

Impact of irrigation development on a less developed farm economy has been far-reaching in dimension and magnitude. The operational features of the irrigation system, particularly canal irrigation, have been examined against the backdrop of the prevailing structural and organisational bases of individual farms. It is important to verify whether regressive impact on income distribution of canal irrigation may be caused by the non-neutrality of economic, social and political institutions rather than the new technology, irrigation, per se. The fact that provision of canal irrigation has resulted in increased cropping intensity is corroborated here by the findings of the case study of the Mayurakshi canal irrigation command area. Based on the different observations, one is invariably led to the conclusion that there has been a departure from the traditional subsistence pattern to a market-oriented commercialized crop pattern. The impact of canal irrigation on cropping pattern is not uniform in all the localized areas (Head Reach, Middle Reach and Tail-End). The canal water is not evenly diffused in all the areas.

Variations in labour use per cropped area may be seen as the combined effects of variations in proximate factors like higher intensity of cropping, diversification of crop pattern, cultivation of labour-intensive crops etc. It has also been observed that the introduction of canal irrigation followed by the adoption of HYV paddy etc. was found instrumental in generating substantial amount of employment opportunities. This hypothesis is also positively affirmed by the findings of the micro-study. It is the farm profit which is the appropriate indicator of economic efficiency.

In this excercise, in estimating the Farm Business Income (FBI) separately with respect to irrigated and unirrigated farms, size-class-wise, it has been observed that, on an average, the Farm Business Income per acre per annum has increased by Rs. 245 predominantly because of provision of irrigation water. So the hypothesis that it is economically more profitable to cultivate on irrigated land stands confirmed.

It is interesting to examine the hypothesis whether the benefits of canal irrigation are largely confined to upper stratum farm families. The gains from canal irrigation are found to be size neutral. There is no evidence to suggest that income and gains are confined to relatively big farmers alone. Rather the higher the size of operational holdings, the lower the differential income gains. However, per acre and mean household income gains are more on relatively smaller farmers. The propagation and diffusion of canal irrigaition system among agricultural households seem to have reduced rather than aggravated income disparities in the study area.

The experience of the past few decades highlights several issues facing Indian irrigation. These are mainly concerning the kinds of irrigation works that should be promoted, the relative emphasis that should be given to quantitative expansion, the means to achieve qualitative improvement, the question of equity in the distribution of benefits from irrigation, the recovery of costs from beneficiaries, pricing and subsidy issues, environmental aspects, stabilising role of irrigation input etc.

It has been observed that it would be appropriate to look upon both large and small works as complementary rather than substitute to each other. It also emerged that the crucial issue is not as much the growth of irrigation potential as the optimal use of the created potential. Another point at issue is the pricing of water. The proposition of supplying irrigation water free of charge holds little ground. This free supply of irrigation water involves broader issues of resource allocation and inter-sectoral and inter-personal income distribution. Under a system of free supply of irrigation water, the non-beneficiaries of both the rural as well as the urban sectors would be required to reimburse the expenses incurred for provision of the irrigation facility to the beneficiaries. And again, another issue is generally raised with regard to the issue of what should be the method of charging irrigation water, i.e., whether water charge should be collected per unit of land irrigated or per unit of water delivered. In fact, economists frequently wish pricing to be considered. Usually, they appear to be under the illusion that it is feasible under the canal irrigation system to charge farmers for water on a volumetric basis. But to our knowledge, there is no canal irrigation system in South Asia where the amount of water issued to an individual farmer is directly measured. This being so, charging volumetrically to provide an incentive for the sparing use of water is not feasible. And finally, the implication of low irrigation rates with added problems of low recovery is that today the water charges collected do not cover even working expenses, not to speak of depreciation charges and return on investment.

Since irrigation is a major source of increased productivity, its distribution between regions and classes is naturally a matter of concern. It is sometimes argued that the relationship between irrigation ratio and the size of holding has been weakened, because large farmers have superior access to well irrigation and even canal irrigation and large farmers have also the capacity to manipulate the distribution of water in their favour. Of course, the first explanation seems plausible, the second explanation is quite tenuous. There is no obvious reason to assume that all advantageous locations are systematically pre-empted by large land holders. Large land holders are found in all reaches of the canal network.

Much of the aspect 'instability' in the case of irrigated farming is traceable to area instability; whereas much of the output instability in the case of rainfed farmers is traceable to yield instability. It is also maintained that stability in yields of all major crops is entirely due to climatic and natural factors and that policy instruments are needed to avert increased instability even in areas with assured irrigaiton. How far irrigation in a state is insulating its farm economy against the ravages of drought is a moot question.

Finally, it is stated that the over-domination of the engineering-centred, capital-intensive project-oriented approach in water distribution is the bane of the crisis. The societal vision is totally missing and the technical solutions have perpetrated. In view of the extraordinary diversity of conditions and complexity of problems in India, solutions have to be derived from a very careful and detailed analysis of specific situations.

In the analysis, it also emerged that there are two major reasons for conjunctive use of water. In the first place, even though the farmers expect that the canals would provide adequate water for the crops they intend to grow, in reality, water supplied from canals is generally found to be inadequate, untimely and thus not dependable. Secondly, the desire to shift the cropping pattern from the traditional crops to modern high-yielding varieties of water-intensive crops has induced farmers in the command area to adopt well irrigation. Again, the need for conjunctive use of ground and surface water is particularly clear given their tendency to mutually offset the ecological imbalance resulting from the use of each source. The farmers face different problems like brackish groundwater acquifers, lack of capital, lack of electricity, uneconomic and scattered land holdings, etc. In the Indian conditions, major versus minor or surface versus groundwater is not substantial issue. Both groundwater and surface irrigation are interlinked hydrologically, giving rise to powerful external effects. The role of canal seepage in the economy of wells established in the regions of low rainfall where natural availability of groundwater is poor is widely appreciated. Dhawan examined the existence of a high rate of return on wells benefiting from canal seepage compared with those not benefiting from it. In a situation in the presence of such externalities, it could not be prudent to talk of taking advantage of comparative costs and returns in planning. Again, being preoccupied exclusively with the development of either source of irrigation can ultimately be costly. All our irrigation capabilities need to be developed, subject, of course, to technical and economic conditions; because even after tapping all exploitable water resources, a large part of cultivation may have to be carried out without the aid of irrigation. Again, today land productivity of canal irrigation is well below the corresponding levels under wells. In the crucial matter of timeliness, an irrigation source which is outside the control of a farmer cannot match one that is wholly under his ownership. With regard to adequacy of irrigation, a protective canal, by the logic of its conception, cannot usher in an era of high crop yields. Even a canal designed for productive farming canal cannot deliver the goods. The actual delivery of water at the final outlet is below the desired value, say, due to successive transit losses. This beneficial role of these water losses from canal seepage has not received adequate attention. In fact, canal water improves the ground water-based agriculture in India.

The management of irrigation encompasses operations of a canal system from the reservoir down to the delivery of water to farmers. In a democratic polity, political and social factors also play their role. Another consideration is the protective nature of irrigation, which is designed for providing insurance against the total failure of crops. Another aspect is the localisation pattern and the issue related to it. The point as to how and when and with what frequency water is to be used is also relevant. In reality, farmers prefer to irrigate during the daytime continuously. Again, actual field conditions may not very often conform with the water supply. This mismatch may generate some problems involving local dynamics. These are some of the critical issues that require to be tackled in managing canal irrigation efficiently.

Maldistribution of water leading to supply of less amount of water to the tertiary level generates problems of spatial inequity. A laissez-faire system of water use tends to develop. This is the end result of a complexity of operations carried out at different points in the irrigation system. The importance of irrigation structures necessary to control and regulate water as per the design of the cropping seasons seems to have been given much importance to prevent unauthorised irrigation. Water use efficiency depends to a great extent upon enforcing the exact cropping pattern. If water-intensive crops like paddy, sugarcane etc. are cultivated in a project area which is designed for light irrigated crops, it becomes difficult to ensure equity in water use. Imbalance in cropping pattern leads to income inequalities among farmers as farmers at the head reaches grow more of commercial crops and get more income and their counterparts at the tail end continue to grow traditional low value crops and get low income. Rotational Water Supply (RWS), as it helps to enforce discipline among farmers to use available water judiciously, is a preferred method of irrigation.

It should be noted that with the introduction of major irrigational works like canals, new administrative hierarchical structures for irrigation management were constituted. These administrative structures were in complete contrast with the more localized and user-participatory management of earlier times. The bureaucratic mechinery which thus got introduced was far removed from the actual issues and therefore much less reliable to the actual need of the users, particularly in terms of maintenance and operation of services.

Making Policies Effective

The benefits from conjunctive use of surface and groundwater are significant. But their extent depends to a great extent upon scientific planning and proper integration of canal operational procedures suitable to soil conditions in the command area. In order to establish a long-term water balance equilibrium over the entire command area, it should be suggested that the canal irrigation use near the head and the main and branch canals should be discouraged. On the contrary, groundwater use is required to be encouraged. This is because this would be an antidote to water-table buildup and so, the most effective way to secure an extensive spread of canal water resources. Achievement of this at the head and along the main and branch canals automatically improves the services to the tail-enders as a canal system. And in this process, it helps to achieve as much equity as possible in the spatial distribution of or access to water.

Where the potentials for conjunctive use of two sources of

irrigation — surface and groundwater — exist, the typical experience has been that in the command area, the utilization of one of the sources declines. In other words, the co-joint availability of two sources becomes competitive, in practice, rather than complementary in use. Again, it follows that the emergence of shallow tube wells, owned and operated by the relatively better-off farmers, in the command area has brought about increasing disuse of canals by some of the landowners. This abstination by these landowners from the use has generated poor maintenance of the watercourses because of lack of co-operation of all the plot holders along a particular watercourse. Maintenance and upkeep of the canal is, hence, found to deteriorate. Again, it has been observed that the small farmers, particularly those dependent on canal irrigation, are generally the victims; they are eventually forced to purchase water from the neighbouring relatively better-off farmers, or to instal their own shallow tube-wells. Both the alternatives – installation of tube-wells and the purchase of water –are practically expensive propositions. Tube-well installation calls for excessive expenditures and a low degree of utilization of capacity for the small farmers, while purchase of water has been found to be more expensive and less dependable. All these bring us to the important question of systematic regulation of ground water utilization. In fact, so far there is virtually a total absence of any direct control over the installation of shallow tube-wells in any command area. Hence, some efforts are required to be made in this case to introduce some kind of control through ground water legislation.

A major canal irrigation network may be viewed as a sum of a number of low capacity irrigation systems. And it is also observed that low capacity irrigation works are, by and large, more efficient; because they require less complicated control of water supply and less co-ordination and the distribution is localized. Now, it is possible to decompose the largescale canal irrigation system into a series of smaller units. In the canal system, it is possible to construct small reservoirs in the command area to which water is to be supplied. The control over quality and timing of water to be released is to be left to the users at the local level. This decentralization of the operation and maintenance of the canal system calls for a systematic planning for designing of the entire scheme at the designing stage itself for building suitable structures.

In the case of major irrigation projects, the major problem is cost recovery. Revenue receipts have generally been poor. There does not seem to exist any uniform principle of charging water rates which vary from state to state. Such variations may be expected because of diverse soil and crop pattern with different socio-economic conditions. But even then, the existing practices are not based upon any comprehensive study of conditions of supply and use of irrigation. The existing rates of public irrigation works are highly subsidized. Its collection, whenever adopted, had fallen short of target. So, in the fitness of things, it is suggested that ways and means of achieving a more rational distribution of costs and structure of rates have to be solved. In this connection, it should also be borne in mind that the water rates should be so fixed as to provide sufficient incentive for productive use of irrigation and at the same time, encourage adoption of progressive technologies.

The sources of irrigation differ substantially across size of land holdings. From the analysis made earlier, one might get the feel that the small and marginal farmers rely more on canal and tanks as a source of irrigation compared to large farmers. Again, subsidies in canal irrigation consists mainly of low water charges levied by states, non-realization of water dues etc. Hence, given a much higher proportion of irrigated land on smaller holdings, total subsidies in canal irrigation may have a bias in favour of small farmers.

Assuming that input subsidies have an elastic consumption and positive yield response, a fall in canal irrigation subsidies may contribute to the reduction in the consumption of inputs and hence, resultant fall in productivity. This may generate adverse distributional impact. Hence, continuation of some subsidies accompanied by better and more effective delivery system and a rationalized input price structure deserves a careful look, especially in the context of its welfare implications (Kumar, 1997).

Existing approach of analysis of canal irrigation treats irrigation predominantly as a 'technological' input in production. Hence, attempts are generally made to analyse extensively the technological, financial and economic problems related to the supply of irrigation. But it is pertinent to note that these technological considerations are required to be placed within the context of socio-economic relations and conditions of production and exchange of the user community. It is not the technological feasibility alone that matters. The creation, maintenance and utilization in totality is contingent upon these production and exchange relations in the user community. Again, efforts should be made to evaluate the interaction between the irrigation technology and the social relations in a region, i.e. on the impact that irrigation technology has on the conditions of production and accumulation of the different categories of users.

The aspect which should be systematically reckoned within the planning of irrigation supplies is the importance of a well- administered and planned maintenance of delivery system. It is observed that even in a canal irrigation system, the provisioning of maintenance is usually left to the private sources. As such, there are inadequacies and uncertainties, because there are very strong externalities of these private investment. In view of this, the work of maintenance and distribution should be ideally co-ordinated through co-operative organization and with local control. This would also enable mobilization of local resources, particularly labour and local findings. Decentralized operation of water systems or construction of smaller-scale irrigation schemes appears to have an advantage in this regard.

To overcome the shortage of water or to rationalize the utilization of water in a canal irrigation system, the water users, particularly those dependent entirely on water supplies, are found to resort to some arrangements like formation of Water Users Organization (WUO) (Hart, 1978). Success of any WUO is not necessarily and always guaranteed. Its successful operation is contingent upon the fulfilment of a set of conditions. For example, the probability of success of a WUO is limited in a caste-class ridden society. Hence, any mechanical attempt to set up WUO in any command area without taking appropriate care of the socio-economic conditions may lead to dysfunctional organisations. This is corroborated by some studies (Pant, 1984; Hart, 1978).

The inequality that persists between small and large farmers in irrigated tract may be attributable to some sort of inequality attained by farmers in the use of non-irrigation inputs which happen to be major sources of increasing crop yield in modern agriculture. Both the quantity and quality of irrigation are a major determinant of fertilizer application, particularly in semi-topic areas like Birbhum district under Mayurakshi canal command area, of low and medium precipitation. Both these factors fundamentally affect the choice of crop varieties and crop pattern. Level of fertilizer application depends largely among others on farmers' own resources position, availability of farm credit at low interest rate, price policy for fertilizer.

Unfortunately, all these factors are normally tilted against small farmers; therefore, they cannot realize as much income benefits from irrigation as do large farmers. In this connection, it should be borne in mind that the access to the non-irrigation inputs may itself be limited to the small cultivators and as found in the United Provinces studied by Whitcombe (1974), the compulsive commercialization under debt-bondage may further deny the petty indebted farmers the advantages of irrigation. Comparing areas dominated by small holdings with those with more even land distribution in Punjab, Jairath (1984) found that agricultural performance is poorer in the former in terms of cropping pattern, physical yield, land use intensity etc. although the areas have comparable level of irrigation in the aggregate.

But rectification of this tilt is largely contingent upon purposive and determined state interventions in certain areas like land reform and consolidation. The surest way to eliminate the access-related tilt is to work towards an equalization of land holdings. Both land distribution and strongly tilting public allocation of irrigation water and other inputs are eminently suited to equalise the income gains from canal irrigation system.

A common property resource like canal irrigation project sometimes get over-used as each one feels if he does not use it, others will. He will utilise water as much as possible, irrespective of the amount he actually requires. When every other user does the same, the canal degrades and eventually, all suffer. Yet no one singly would act otherwise. This 'Tragedy of Commons' describes here the situation as over-use or injudicious use of water by some users.

In this context, the damage thus caused to environmental quality is external to the farmers' profit and loss account. A farmer who mis-uses or abuses water could pollute the soil-quality and deteriorate the level of productivity of land of the area. But that particular farmer does not pay for the damage. In a sense, environment belongs to all and therefore, it belongs to none. No one cares for it unless all agree to care for it. Environmental problems arise mainly due to externalities.

Many traditional societies have evolved over time codes of conduct which result in a sustainable use of these resources. These, of course, break down under the pressure of growing population and onslaughts of commercialization. It is suggested that what is needed is just well-defined appropriate property rights which should be enforceable and secure if they are to lead to sustainable resource use. These policies may internalise the externalities, so that environmental costs of one's actions are to be borne by the individuals concerned. Tax policy may also be a tool to do this.

Water use efficiency depends largely on enforcing cropping pattern envisaged in the project design. One should know whether farmers are really aware of the recommended crops. Farmers, otherwise, may be guided by market forces. Quite often, farmers appear to be locked into their own variety of 'Tragedy of Commons'. The abundant issue of water through gravity flow system, and consequent flooding remove any option for a farmer to grow anything but paddy. It is true, a possibility of collective change to a more productive, more profitable and more water sparing cropping system is practically difficult. What is desirable is to try a potentially feasible cropping pattern on a limited area and then to promote collective decisions to change.

There is also a need to regulate cropping pattern through controlled water distribution system. Demand-based irrigation system may be introduced. Farmers are to inform in advance of their own requirements to the irrigation officials. This helps to plan water budgeting and allocation.

Irrigation tariff should be charged on volumetric basis. There should be adequate judicial power of the irrigation officials to control and regulate water use and distribution to ensure discipline.

Attempt should be made to consolidate the land holdings to rationalize the holdings to overcome fragments. Sometimes resistance is offered by the better-off farmers. Because they see this programme as detrimental to their irrigational interests. Owning plots of land in different areas is considered as an insurance against risk of water shortage which may be locally concentrated. Again, if the plots are scattered, they continue to keep their hold on water distribution and other related social dynamics. Small farmers are also apprehensive of losing the good quality of land. Small farmers have small and tiny and scattered plots of land. So, the issue of consolidation does not arise in this case. Ultimate success depends largely upon the political will to implement it.

Attitude towards water distribution in the existing projects needs to be corrected through deliberate public policy. Under Rotational Water System (RWS), wastage of water is practically nil and water supply is reliable and timely. Under RWS system, rotational use of water is possible. So, RWS instead of Continuous Flow System (CFS) should be introduced.

Again, the present 'area' concept may be replaced by 'time' concept. Instead of notifying that a farmer gets water for 'x' acres, it should be notified as 'x' hours. These water hours may be allocated on pro-rata basis of land ownership. On the basis of water-hour entitlement, farmers can decide the type of crops they should grow and the extent of area to be irrigated.

Kharif and Rabi zones should be fixed under separate distributories keeping control at the main canal. Conjunctive use of water should be encouraged.

Construction of farm ponds should be encouraged at appropriate locations to store water. The very concept of a farm pond and its operational mechanism is to be made clear to the farmers.

Influential and big farmers normally resort to breaking the structures to draw more water to complete irrigation during the daytime. Officials or persons concerned should have full authority and judicial power to tackle irrigation offences.

In the light of experience regarding inadequate, untimely and non-dependable water supplies from canal and the desire for growing modern high-yielding varieties of commercial crops like sugarcane etc. the farmers should be induced to adopt well irrigation in canal command area. Efforts may also be made to consolidate lands to provide community wells to provide facilities to small and marginal farms which are uneconomic. At the same time, care should be taken to regulate the unorganized proliferation of wells.

There is a need for lining the distributories and also outlets to save water, whenever required. But it should be borne in mind that geophysical features of the region and the specific regions are required to be studied before advancing any concrete proposal in this respect. Lining of channels may be an uneconomic proposition if the soils have high water retentivity.

Farmers' participation in irrigation management should become an integral part of project administration. Irrigation Committees at the outlet and distributory levels have to be formed with proper representation to tail-end and other farmers belonging to weaker sections.

There is a need to intensify agricultural extension activities in the command area to educate farmers on water use and management practices.

Maintenance of field irrigation channels should be made obligatory to the farmers. It should be made a rule that all farmers should desilt and clean the field channels before releasing water.

The environmental aspects of water resource development, particularly canal irrigation network may be explained in terms of –

(a) the direct effect that construction of large reservoirs has or the immediate environment (like the submergence of economicallly and ecologically valuable forests as well as the displacement of people living in the area),
(b) damaging cultivated land by indiscriminate irrigation,
(c) the high rates at which reservoirs are silting up, and
(d) fear of over-exploitation of groundwater.

Measures to check deforestation of upper catchment and then to restore forest cover to an appropriate density should be an integral part of planning for irrigation projects. This should be done on a watershed basis. It is again perferable to rely on such solutions as a number of small, local storages or groundwater development that do not damage the forests and to minimize human problems arising from displacement. Relatively large storages that impound the surplus flows during the short monsoon season are a technical imperative. One could think of fever large storages combined with and linked to smaller one's. This possibility is required to be examined in each specific context.

It is true that extreme position of the environmentalists for the proper rehabilitation of the displaced persons stens from their frustration. The answer, of course, does not be in environmental fundamentalism. Preventive/Corrective actions should be an integral part of the design and implementation of the projects. It is essential to evolve a procedure by which the techno-economic evaluation of the projects with a displacement effect/environmental impact are required to be made available to the people in the areas to be affected by the project to elicit their reactions.

There is a strong viewpoint that water charges should be a function of volume and season. The Report of the Committee on Pricing of Irrigation Water (1992) suggested that the existing system of assessment should be rationalized and simplified based on crop-wise irrigated area on individual basis after estimating the relative water consumption per hectare irrigated in different seasons. It also suggested that the assessment and collection functions should be entrusted to the Irrigation Department of the State after empowering it with the Revenue Recovery Laws. The Committee was also of the view that the country should move over progressively from management through government bureaucracy to management by user farmers who would take over maintenance, management of water allocations and collection of water rates for a group of outlets serving at least each village.

An expert committee on pricing of irrigation Waters (GOI, 1992) recommended as increase from existing levels of Rs. 50 per hectare to Rs. 310 per hectare. This would cover operation and maintenance costs and 1 per cent of capital costs. The committee also recommended that the irrigation commands should limit themselves to wholesale distribution of water with volumetric pricing to farmer groups.

These arrangements may reduce the wasteful use of water, contribute to better allocation of water within irrigation commands, provide funds for improved operation and maintenance and partly reduce the rent-seeking activity generated by the present administrative methods by which water allocated in most systems.

But this institution of arrangements may not help allocate water between different uses. It is suggested that efforts should be made to create conditions which would allow the existence of efficient markets in tradable water rights. If rights to the delivery of water can be freely bought and sold, farmers with new crops or in new areas will be able to obtain water provided they are willing to pay more than its value to existing users and established users will take account of its sale value in deciding on what and how much to produce. In this way, there is a great potential for mitigating some of the pervasive problems of Indian irrigation commands, for examples, the tail-ender problem. (Pursell and Gulati, 1996)

Again, it should be maintained that the receipts from water charges should go directly to the irrigation authority. It is observed that at present, a major problem is that water charges go to the general state revenue and so there is no direct connection between improved efficiency in collecting water rates and the availability of funds to improve the operation of the irrigation systems. This undermines both the motive of farmers to pay water charges and the interest of the irrigation command staff to collect them.

Since the command area is fixed in size and defined in terms of agro-climatic and geohydrological features, the productivity of water can be affected by choice of appropriate interventions in :

(i) release of water into different sections of the irrigation network;
(ii) choice of appropriate technology for water distribution system, lining of canals, control structures etc;
(iii) institutional mechanism for effective enforcement of water distribution rules–administrative component and the role of farmers' organizations;
(iv) monitoring and concurrent evaluation of performance of the system;
(v) improved on-farm efficiency of water-use and choice of crops;
(vi) augmenting water supplies with the ground water and rain water; and
(vii) other economic interventions in the form of incentives for higher productivity.

A canal is a channelled flow of water operating on the basis of a gradient. The source of water may be a river or a dam or a reservoir at the point of take-off. The total discharge at the canal head is subject to variations which are not always predictable because of variability in the rainfall etc. Another major problem is the transmission of water to the fields through a network of canal. The efficiency of water transmission is critically influenced by the installation of canal structures, their maintenance and organization of water distributories. It is true that violation of technical requirements of a particular gradient in the watercourse may impair the amount of water actually reaching the field. Any attempt to revamp that terminal delivery system in a canal project with the help of a standard set of policy prescriptions without adequate recognition of the local conditions may be frustrating (Wade, 1992; Sengupta, 1982). Again, the productive efficiency does not exclusively depend upon technical capabilities and characteristics of the system like creation or installation, finance, maintenance and utilization, but also on the particular socio-economic features and conditions of production and exchange relations prevalent in the community of users (Bharadwaj, 1990).

ANNEXURES

ANNEXURE 1

THE MAYURAKSHI RESERVOIR PROJECT
SALIENT FEATURES

After years of study, Mayurakshi Reservoir Project was drawn up. The Mayurakshi, a hill stream, originates from the uplands of Santal Pargana in Bihar and falls into the Bhagirathi after a course of 150 miles. Though a small stream, it is the largest of the rivolets that run through the districts and its total annual flow, even in a year of drought, is sufficient to irrigate the major part of the districts. But most of the annual discharge passes down this river in swift and sudden floods during the monsoon. Except in flood, the river does not carry any appreciable volume of water. A storage reservoir was, therefore, considered to be the only means to hold the water and supply it to the fields. Fortunately, at Massanjore, in Santhal Pargana, Bihar, where the river flows through a narrow valley, an ideal site for the dam was found. Based on supply of water from a storage reservoir to be constructed there, a multipurpose scheme was drawn up for irrigation of power and flood control, with the first, of course, as the principal objective.The irrigation work ultimately is classed as 'productive'.

The Project, essentially, consists of a storage reservoir at Massanjore of gross capacity of 5,00,000 acres ft. and five pick-up barrages, one each at Tilpara, Boocha, Baidara, Kadisala and Kultor over the rivers Mayurakshi, Dwarka, Brahmani, Bakreswar and Kopai respectively. The Project aims at irrigating net command area of 5,60,000 acres and 50,000 acres of land for kharif and rabi crops respectively in the districts of Birbhum, Murshidabad and Burdwan. In addition to above, a total area of about 20,000 acres in the territory of Bihar state is supplied with irrigation water through a canal directly from the Massanjore Dam. The main barrage and headworks at Tilpara were completed in June,1951 and irrigation for kharif crops was allowed from 1951.

The whole of the command area of the Project within West Bengal has been divided into Blocks, as detailed below:

Block A.	Brahmani-Pagla Basin	74,000 acres
Block B.	Dwarka-Brahmani Basin	77,000 acres
Block C.	Mayurakshi-Dwarka Basin	1,37,750 acres
Block D.	Mayurakshi-Brahmani Basin	76,750 acres
Block E.	Bakreswar- Kopai Basin	45,000 acres
Block F.	Kopai-Ajoy Basin	1,53, 500 acres
	Total	5,60,000 acres

I. Salient features of the Massanjore Dam

(a) The Dam is located in between two hillocks at Massanjore in Bihar and across the river Mayurakshi at about 60 miles from its source. It is a masonry dam 2,170 feet long with a roadway on the top. The storage reservoir has an area of 27 square miles.

Massanjore Dam:

(i)	Height – 155 ft. above the deepest foundation level	
(ii)	Level at maximum storage	398.00 ft.
(iii)	Catchment area	718 sq. miles
(iv)	Submerged area	27 sq. miles
(v)	Number of floodgates	21 nos.
(vi)	Expected life of the dam	100 years

II. Salient features of the Tilpara Barrage

The Barrage is located over the river Mayurakshi at Tilpara about 20 miles downstream of the Massanjore Dam. It has seven weir bays and eight undersluice bays of 60 ft. wide each, having a combined discharging capacity of 2,90,000 cusecs (max).

(i)	Total length	1,013 ft.
(ii)	Catchment area: —controlled	718 sq. miles
	—uncontrolled	521 sq. miles
(iii)	Storage capacity	7, 350 Ac. ft. above sill level of head regulators
(iv)	Discharging capacity of the barrage	2,90,000 cusecs

III. Salient features of Dwarka Barrage

(i)	Length	275 ft.

(ii) Catchment area — 17 sq. miles

(iii) Discharge capacity of the barrage — 40,000 cusecs

(iv) Discharge capacity of head regulators — 1695 cusecs

IV. Salient features of Brahmani barrage

(i) Length — 485ft.

(ii) Catchment area — 259 sq miles

(iii) Discharge capacity of the barrage — 69,000 cusecs

(iv) Discharge capacity of the head regulators — 825 cusecs

V. Salient features of Kopai Barrage

(i) Length — 205ft.

(ii) Catchment area — 82 sq. miles

(iii) Discharge capacity of the barrage — 26,500 cusecs

(iv) Discharge capacity of head regulators — 2,410 cusecs

Particulars of Mayurakshi Reservoir Project Canal System

Name of Main Canals	*Length (km)*	*Capacity at off-take point (Cusecs)*	*Command Area (Acres)*
(i)	*(ii)*	*(iii)*	*(iv)*
Mayurakshi-Bakreswar Main Canal	22.91	3.500	2,75,250
Bakreswar-Kopai Main Canal	11.54	2,410	1,98,500
Kopai South Main Canal	79.28	1,850	1,53,500
Mayurakshi-Dwarka Main Canal	16.88	3,500	2,84,700
Dwarka-Brahmani Main Canal	38.22	1,695	1,51,000
Brahmani North Main Canal	47.88	825	74,000

Note : Total length of branch canals: 147.07 km.

There are four Branch Canals taking off from the Main Canals, particulars of which are given below :

Name of Main Canals	*Length (km)*	*Capacity at off-take point (Causcs)*	*Command Area (Acres)*
(i)	*(ii)*	*(iii)*	*(iv)*
Mayurakshi-Bakreswar Branch Canal	47.05	770	69,300

(Contd.)

(i)	(ii)	(iii)	(iv)
Mayurakshi-Dwarka Branch Canal	42.43	1,330	1,19,700
Bakreswar-Kopai Branch Canal	30.54	570	51,300
Dwarka-Bakreswar Branch Canal	27.05	480	43,200

Notes : Total length of branch Canals: 147.07 km.
Total Command Area: 2,83,500 Acres

In addition to Main and Branch Canals, there are distributories, minors, sub-minors and a good number of outlets throughout the Command area of the Project.

In all fairness, the Project is responsible for giving the districts a fresh lease of prosperous life by the establishment of irrigation dam and its channels. It has injected a stimulus and a welcome stress and strain in the social and economic spheres of the districts and dissolves their atrophy.

ANNEXURE 2

FUNCTIONAL ANALYSIS

An attempt has been made to examine through production function analysis the impact of irrigation on crop productivity. The conventional neo-classical production function of the Cobb-Douglas type [1] has been chosen for some reasons. This approach has been adopted to examine the input-output relationship separately for two types of farms, viz., irrigated and unirrigated. This function has been fitted to the data to study the relationship between independent variables such as human labour, bullock labour, manure and fertilizer and irrigation charges and the dependent variable, that is, output per hectare for irrigated farms. For unirrigated farms, the irrigation charges have not been considered for study of the relationship.

For this study on two sets of farms, such as irrigated and unirrigated, the regression analysis has been carried out separately with the following selected variables and form of function : $Y = Ax_1^{\alpha 1} x_2^{\alpha 2} x_3^{\alpha 3} x_4^{\alpha 4}$

Where:

Y = gross returns from all crops in rupees

X_1 = human labour used on farms per hectare in rupees

X_2 = bullock labour used on farms per hectare in rupees

X_3 = manures and fertilizers used on farms per hectare in rupees

X_4 =Irrigation charges per hectare per annum in rupees

A is constant term or scale parameter

α_1 α_2 α_3 and α_4 are the partial elasticites of different inputs used.

Before discussing the results, a brief reference to the problems associated with the concepts and estimation procedure adopted in this study is in order. In the first place, it is assumed that crops composition of the total production of the various farms is the same i.e., there are not product quality variations. Secondly, labour is measured in days. There are some unsatisfactory points about labour. One, quality differences of labour are not taken into account. Two, female and child labour are treated at par with male labour. Three, labour is employed on both permanent and casual bases. Nevertheless, inclusion of an unsatisfactory defined variable like labour in this case, is better than to exclude altogether a relevant variable.[2] Secondly, inputs like fertilizer, pesticides, weedicides etc. operate through the soil and are viewed as improving the plant productivity. These inputs account for a significant proportion of the total expenditure. Although there is likely a high degree of correlation among these inputs, all variables are considered here because of their obvious importance in increasing production. All these inputs are treated as a single input.[3] Thirdly, inclusion of some variables which may be irrelevant reduces the degrees of freedom and may increase the possibility of multi-collinearity, thus leading to imprecise estimates of the parameters. Of course, the multi-collinearity does not pose any problem unless it is sufficiently high relative to the overall degree of multiple correlation among all variables. And finally, some economists like Griliches, Kmenta, Mundlak have justified the use of single equation model for agricultural production function. They argue that since inputs in agriculture are largely predetermined and there is a considerable lag in production and error is largely weather-determined, simultaneous bias will be minimum if the production function is well-specified.[4]

Results and Discussion

The production elasticities and their standard errors and computed student's for irrigated as well as unirrigated areas and the corresponding co-efficients of multiple determination (R^2) have been presented in A-2.1.

Table A-2.1 : Regression co-efficients, Standard Errors and co-efficients of Multiple Determination for Irrigated and Unirrigated Area in Mayurakshi Canal Command Area.

	Irrigated Area		*Unirrigated Area*	
Variable	*Regression (α_i) Coefficient*	*Standard Error*	*Regression (α_i) Coefficient*	*Standard Error*
(1)	(2)	(3)	(4)	(5)
Human Labour	0.481 (32.32)	0.012	0.262 (4.12)	0.02
Bullock Labour	0.023 (2.8)	0.03	0.211 (3.8)	0.12

(Contd.)

(1)	(2)	(3)	(4)	(5)
Manure and Fertilizer	0.652 (74.8)	0.001	0.181 (016.5)	0.01
Irrigation Charges	0.011 (18.5)	0.003	—	—

R^2 = 88 per cent
Number of observations : 98

R^2 = 81 per cent
Number of observations : 64

Note : Figures in parentheses indicate statistic.

The Table reveals that for the irrigated area, the elasticities of production for human labour, bullock labour, manure and fertilizer and irrigation charges being 0.481, 0.023, 0.652 and 0.011 respectively showed significant impact on the value of the gross return per hectare. For the unirrigated area, the production elasticities for human labour and manure and fertilizer were respectively 0.262, 0.211 and 0.181 and were found to be significant both at .05 and .01 levels. It is also clear from the exercise that all the inputs like human labour, bullock labour, manure and fertilizer and irrigation charges were responsible for explaining 88 per cent of the variation in the value of output per hectare in irrigated area. On the other hand, inputs like human labour, bullock labour, manure and fertilizer explained 81 per cent of variation in the value of output per hectare.

The marginal value products (at the geometric mean level) for all the inputs both in the irrigated and unirrigated areas shown in Table A-2.2, state that in the irrigated area, the marginal value products of human labour, manure and fertilizer and irrigation are significantly higher than their prices. This suggests that there is lower than optimum level of human labour, manure, fertilizer and irrigation. Gross return per hectare would be increased by increasing their levels. This, in turn, suggests that expansion of irrigation is a pre-requisite for increase in agricultural production. The difference between marginal value products of bullock labour and its production on unit price was found to be non-significant. This indicates the optimal use of the input in the irrigated area. But in the case of the unirrigated area, the differences between the prices of bullock labour, manure and fertilizer and the marginal value products of these inputs were again not found significant; therefore, these inputs were also optimally utilized. But the differences between the marginal value products of human labour and its price were found to be significant. This suggests that in the unirrigated area, human labour is utilized lower than optimum.

Table A-2.2 : Comparison of Marginal Value Products of Inputs for irrigated and Unirrigated Canal Command Area

	Irrigated Area	*Unirrigated Area*
I. Human Labour		
Marginal value product at		
Geometric Mean	1.58	0.72
Price per unit	1.00	1.00
Difference	0.58	0.28
Standard Error	0.03	0.12
II Bullack Labour		
Marginal value product at		
Geometric Mean	0.82	1.31
Price per unit	1.00	1.00
Difference	0.18	0.31
Standard Error	0.18	0.61
III. Manure and Fertilizer		
Marginal value product at		
Geometric Mean	3.18	1.28
Price per unit	1.00	1.00
Difference	2.18	0.28
Standard Error	0.34	0.11
IV. Irrigation Charge		
Marginal value product at		
Geometric Mean	3.11	—
Price per unit	1.00	—
Difference	2.11	—
Standard Error	0.26	—

NOTES

1. Cobb-Douglas production function allows diminishing marginal productivities to each input factor as well as increasing or diminshing returns to scale. It assumes a constant elasticity of production over the entire range of inputs. This type of function can be expressed linear in log and is easy to fit by the method of least squares.

2. Griliches has shown that omission of such a variable causes bias in the analysis – Griliches, Zvi: 'Specification Bias in Estimates of Production Function' in *Journal of Farm Economics*, February, 1957, Vol. XXXIV, No. 1, pp. 12-43.

3. It should be noted that aggregation over inputs may cause a bias in the estimate. Of course, the bias, if any, may be reduced if arithmetic sum in aggregating over input is used. We have, therefore, used arithmetic instead of geometric sums in our estimates.

4. For detailed analysis, see 'Economics of Technical Change in wheat Production in the Indian Punjab' by Surajit Sidhu in *American Journal of Agricultural Economics*, Vol. 56, No. 7, May 2, 1974.

ANNEXURE-3

CANAL IRRIGATION—ITS IMPACT ON FACTOR USE AND PRODUCTION

There is no denying the fact that increase in production is to come mainly through multiple cropping and increase in productivity in land which involves the provision of irrigation facilities.

With this background in view, attempt is made here to examine whether the irrigated farms have higher productivity than the rainfed farms, comparing them for the same crop and season (kharif paddy) and also to explain the nature of technological change.

Commensurate with these objectives, data from a purposive multistage stratified sample of a total of 70 farms following similar cultural practices have been collected from two villages in Birbhum, one is in irrigated region and the other in rainfed tract for the year 1995-96. Of these 70 farms, forty are in the irrigated regions and the rest thirty are in rainfed areas. Birbhum is the district over which Mayurakshi canal flows. The villagers were purposively selected, keeping in view the transport facilities for survey work and that they should be contiguous to each other to keep the uniformity in soil conditions.

For determining whether there is any structural break in production relations governed by irrigation introduced (in paddy production) in the study, Chow test (Chow, 1960) has been employed. For this purpose, the following production functions are specified :

$$\log Y_i = \log a_0 + a_1 \log N + a_2 \log K + U_1 \qquad ....(1)$$

$$\log Y_{ui} = \log b_0 + b_1 \log N + b_2 \log K + U_2 \qquad ....(2)$$

$$\log Y_p = \log c_0 + c_1 \log N + c_2 \log K + U_3 \qquad ....(3)$$

$$\log Y_p = \log d_0 + d_1 \log N + d_2 \log K + d_3 D + U_4 \qquad ....(4)$$

where Y_i = yield per hectare in quantities

N = labour input in mandays per hectare

K = current value of fertilizer, farm-produced manure and new seed per hectare

D = dummy variable which takes the value 'one' for irrigated farms and 'zero' for unirrigated farms

i, u_i and p = irrigated, unirrigated and pooled production functions respectively.

In this connection, it should be stated that female labour engaged in production has been converted into man-equivalent by treating one female equal to .75 man. And 'capital' (K input) practically includes, among others, depreciation charges, repairing charges of implements, interest expenses in addition to fertilizer etc. mentioned earlier.

The presence of structural break in paddy production for both the groups has been identified by comparing separate regression equations (1) and (2) with the pooled regression equation (3) with the help of Chow test. From the Chow test, we can only infer whether the function has changed. To decide which coefficient has changed, we use dummy variable. To test the hypothesis that the irrigation has affected slope, we include in the function the factor D, an additional regressor in Equation (4) and then test the significance *of* d_3.

The results of the above regression estimated by OLS method are presented in Table A-3.1.

Table A-3.1 : Per Hectare Production Functions for the Irrigated and Unirrigated Farm

	No. of observations	*Constant log A*	*Coefficients of* Labour (N)	Capital (K)	Dummy (D)	R^2	*Residual sum of squares* (e_i^2)
	(i)	(ii)	(iii)	(iv)	(v)	(vi)	(vii)
Irrigated (1)	40	-3.14	.32 (.20)*	.34 (.11)**		.81	6.8(e_i^2)
Unirrigated (2)	30	-2.5	.38	.19		.76	18.6
Pooled (3)	70	-2.1	.48* (.27)	.31 (.18)		.85	30.2(e_i^2)
Pooled (with Dummy Variable) (4)	70	-2.9 (.18)*	.35 (.16)*	.38 (.07)*	.14	.80	32.5

*Indicates significant at 5 per cent.

**Indicates significant at 1 per cent.

The calculated value of F^* was found to be 4.25; whereas, the theoretical value of F at 95 per cent level of significance with $V_i = K = 3$ and $V_2 = n_1 + n_2 - 2K = 64$ degrees of freedom is 2.8. Thus $F^* > F_{.0.05}$ and we reject the null hypothesis (H_0): the two relationships do not differ significantly and conclude that the relationship is changing from sample to sample; the difference is not to be attributed to chance. Once the presence of structural break is established, we examine whether the shift in production function has been due to irrigation. In this exercise, d_3, the coefficient of D, is found statistically significant at 1 per cent level. This exercise lends credence to the fact that the introduction of irrigation caused a structural shift of the original production function. Paddy responds consistently better in the irrigated farms. This is presumably because of a better water supply and higher amount of fertilizer use. It is true that risk of crop failure is quite high in rainfed areas. Adequate availability and better control of water motivate the risk-averting farmers to increase the expenditure on the modern inputs like HYV seeds, fertilizers, chemicals etc.

The partial elasticities of the production functions are the factor shares in total output. We can utilize these for examining the technical change-biases with many factors of production. The co-efficients obtained are presented below in Table. A-3.2. The partial elasticities of the production functions reveal that the share of labour decreases substantially when we move from unirrigated farms to irrigated farms. This may be due to the fact that a given amount of paddy could be produced with less amount of labour under irrigated farms. Increase in use of 'capital' input due to irrigation appears to be much higher than that in the use of 'labour' input.

Table A-3.2 : Estimates of Production Elasticities

Inputs	*Unirrigated farms*	*Irrigated farms*
Constant (A)	–2.5	–3.14
Labour (N)	– .38	+ .32
Capital (K)	+ .19	+ .34

According to Hicks, technical change is neutral (Hicks' neutral) if the ratio of marginal products of labour and capital in a two-factor model like ours remains constant (Hicks, 1994). There will be bias in technical change in a given capital output ratio; labour – or capital-using—if marginal rate of substitution of capital for labour increases or decreases. Binswanger, however, has developed a model slightly different from that of Hicks to explain biases in technical changes (Binswanger, 1974). He sought to examine biases in terms of factor's share in total cost. Technical change is labour-saving, labour-neutral or labour-using depending on whether labour share in total cost decreases, remains constant or increases at constant factor prices. In other words, technical change can be measured on the basis of the changes in the output elasticities. Symbolically, in discrete terms, in our case, the nature of technical change can be stated as follows:

$$X_i = \frac{(a_i)_I - (a_i)_{ui}}{(a_i)_{ui}}$$

where a_i is the output elasticity of the i-th factor and the subscripts i and ui are irrigated and unirrigated technology According to Hicks' definition, if $x_i > 0$, the technical change is the i-th input-using; if $x_i < 0$, the technical change is the i-th input saving and if $x_i = 0$, the i-th input is neutral. Table A-3.3 reveals that irrigation technology is biased in favour of capital and against labour. Of course, our results here indicate only the relative position of the factors' share in total income.

Table A-3.3 : Nature of Technical Bias

	Factor	*Proportionats change in output elasticity*	*Nature of technical bias*
Unirrigated	Labour	.38 - .32 = - 0.06	Labour Saving
Vs			
Irrigated	Capital	.19 - .34 = + .15	Capital Saving

But it should be borne in mind that even though relative income shares between irrigated and unirrigated remain unchanged, absolute income share to each factor may change. This change in absolute factor shares gives a better understanding of the income distribution situation. We can obtain absolute share of each factor by multiplying total income by the factor shares.

It is evident from Table A-3.4 that both labour and capital gain under irrigated farming technology in absolute terms. The absolute as well as percentage gain is higher for capital than for labour. The rise in absolute share of labour may be due to rise in employment per hectare. Gain in capital is also significant which is a major cost component in terms of fertilizers, pesticides etc.

Table A-3.4 : Percentage Change in Absolute Factor Shares

Factors	*Absolute factor shares per hectare (Rs.)*		*Percentage change*
I	*II*		*III*
	Irrigated	*Unirrigated*	
Labour (N)	235	295	25.5
Capital (K)	224	310	38.4

NOTE

For conducting Chow-test, we have to obtain F^* ratio. F^* ratio is obtained as follows:

$$F^* = \frac{[\Sigma e_p^2 - (\Sigma e_1^2 + \Sigma e_2^2)]/k}{(\Sigma e_1^2 + \Sigma e_2^2)]/n_1 + n_2 - k}$$

with $V_1 = k$ and $V_2 = n_1 + n_2 - k$ degrees of freedom

where e_1^2, e_2^2 and e_p^2 are the squared sums of the unexplained or residual variations of samples I, II and pooled respectively; n_1 and n_2 are the sample size and k the number of parameters. If $F^* > F$, we reject the null hypothesis, i.e. the two samples differ significantly.

ANNEXURE-4

A NOTE ON CO-OPERATION IN A HYDRAULIC ECONOMY

Water reform is quite indispensable in the sense of building broadly-based community institutions. It is at least as important as land reform. Instead of irrigation bureaucracy, the local community-level issues of coperation are often cited as the key to substantially improving the existing levels of utilization of potential of irrigation. It is sometimes maintained that these institutions can perform different functions : they attempt to pool resources and efforts in constructing and maintaining field channels at the outlet level on the main canal systems to regulate water allocation and monitor violations, to repair, maintain and control water allocation, to control ground water over-exploitation etc.

But the history of local community-level organization is rather mixed. There are certain documented examples of successful local-level co-operation in water management (Wade, 1987; Ostrom, 1990; Tang, 1991). On the other hand, there are different examples of failure of such co-operation, often leading to 'anarchy syndrome' in the appropriation of water (Hart, 1978). In this context, let us try to understand the conditions working for and against sustainability of local-level co-operation (as opposed to government-mandated water-user associations) in situations of economic and social interdependence. We shall try to draw some lessons from the literature on co-operation in repeated game theory in the light of some of the pragmatic insights one can glean from the field studies on the subject.

Pervasive pessimism on co-operation is quite evident from the noted work of Mancur Olson on political economy or that of Hardin on 'the Tragedy of the Commons'. Defection or no defection is the dominant strategy of each player, no matter what the other players do, as cited in 'The Prisoner's Dilemma game'. On the other hand, Taylor (1987) suggested that costs and benefits of collective action on common pool resources like water are often of a kind that is much more favourable to the possibility of co-operation than the 'The Prisoner's Dilemma game'.

It is often suggested that in the case of many vital common pool resources like water, the consequences of defection, on the part of both the agents, may be so bad that either of them would rather do the work himself even if the other does not take care of. The powerful people in the village may resort to this. Some work like field channel maintenance may not necessarily always go by default. Again, it has been observed in the field studies that in some cases, each farmer co-operates when the other does; but defects when the other defects. This is a case of 'Assurance Game' – nobody wants to be 'suckered'. In fact, one tends to be co-operative when the others are: this is the point which 'The Prisoner's Dilemma game' did not capture.

Even in the repeated game theory literature, the passimism in the case of 'The Prisoner's Dilemma game' has been modified in the sense that co-operative equilibria can be spontaneously sustained by the long -run interests of the foresighted self-interested individuals. The possibility of co-operation may depend on the future pay-offs, not being discounted too heavily.

Seabright (1993) has a model of 'Trust'. It can be stated here that the theoretical models suggest a large number of equilibria. Players use observed past behaviour as a guide in their choice. Here what degree of trust the players have in one another is a crucial point. Seabright stated that each player's trustworthiness is confirmed by subsequent behaviour. And there is a tendency for co-operative behaviour to enhance the prospects of successful further co-operation. Trust matters for the possibility of co-operation. Conditions under which co-operation survives have also been examined in evolutionary biology. Smith (1982) in his evolutionary stability maintains that there is a pattern of behaviour such that any small number of people who deviate from it will do less well than the others; no 'mutant' strategy can invade the population.

In short, the Game-theoretic models provide an important insight into the sustainability of co-operation among self interested agents in a situation of interdependence in the management of common pool resources like water. But at the same time, it is important to recognize that these models do not take care of some of the salient issues in real-world co-operation. These models are too mechanical and rigid. These models cannot explain the impact of ongoing interactions among the players in the updating and contingent modifications of the rules of the game, changes in the pay-off matrix etc. These models also ignore the importance of group dynamics which through deliberations may bring about endogenous preference changes and reorientation of values in a commuinty, the importance of leaders acting as catalytic agent in initiating co-operation, etc. Amartya Sen (Sen,1987) has pointed out that the language of 'The Prisoner's Dilemma game' theory does not permit to discuss the behaviour pattern in adapting to the recognition of mutual interdependence; it does not go beyond 'self-goal choice,' as Sen maintained. This 'social norm' is not valid in a large water community.

In this connection, it is interesting to note that in different field studies including ours, people are found to be motivated by the norm of fairness. People do not always desire to free-ride on the co-operation of others. Sociologists like Coleman (1990) state as to how individuals in small communities incur high costs for the welfare of their team. Of course, such team spirit may not be applicable in faction-ridden villages. Intra-village conflicts may hamper the process of internalization of different social norms. The internalization of social norms again depends upon factors like the degree to which the individual identifies with the group, the degree to which the norms are found legitimate, etc.

Co-operation is found quite to work well in small groups with similarity of hydrologic needs. Social sanctions are easy to implement through different mechanisms. Migration and mobility work against co-operation. Exit option reduces the possibility of effective co-operation. Success of co-operative organization in water management depends crucially also upon the hydrologic circumstances and the extent of ecological stress.

An increase in the disparity in private benefits from common property resources management can lead to a situation in which some parties may lose from co-operation. Intra-group heterogeneity in pay-off has negative consequences for the sustainability of co-operative system. In short, social heterogeneity in pay-off, intra-village conflicts, inequality in resource endowment – all impede collective action in water management. The degree of trust or mutual confidence in an inegalitarian environment may be low. Field studies confirm all these theoretical expectations. A low degree of variance of the average annual income among irrigation tends to be associated with a high degree of involvement and commitment to maintenance. Jayaraman (1981) and Easter and Palanisami (1986) in their case studies corroborated the idea that the relatively egalitarian structure of the community is an important factor in the farmers coming together to form water-user organization. Again, the distribution of control over land in relation to proximity to the outlets strongly influences the nature of response of the farmers to co-operation. Sometimes, land fragmentation may help co-operation, because some of the big farmers may have fragmented land in a disadvantaged location. This may induce the farmers to enter into co-operation.

It should be noted that sometimes in some areas in the past, unequal patron-client system prevailed. In this situation, the powerful enjoyed some disproportionate benefits from the system of co-operation. They used to enforce the rules of game and hence gave leadership. But recently because of advent of participatory politics and social changes, legitimacy of this traditional system has significantly eroded. Different rural communities are now in the difficult transition period between the decline of the traditional co-operative institutions and the new modern water-user institutions.

ANNEXURE 5

IRRIGATION MANAGEMENT AND PERFORMANCE OBJECTIVES

Irrigation management is the process by which water is manipulated and used in the production. A widely prevalent belief is that the needs of irrigation management are dictated by the goals of higher agricultural productivity. And it is also sometimes maintained that these goals may relate to productivity per unit of water or land depending upon whether water or land is the constraint. Generally, in public irrigation systems, the farmer's aim is to optimize the productivity per unit of land; while the irrigation agency's aim is to optimize the productivity per unit of water. Consequently, the conflict between private and social goals arises in irrigation management (Gopinath and Kalro, 1991). In most developing countries, irrigation and its effective management are also perceived as instruments to improve employment opportunities, incomes, social justice, equity in social and economic distribution of assets and environmental conditions.

The purpose of an irrigation system is to serve and sustain an agricultural system in a timely and adequate manner so that water, together with other inputs and requisite infrastructure, creates favourable conditions for achieving high levels of crop production and productivity. The components of a surface irrigation system that contribute to its overall performance are: a catchment contributing to run-off; a storage reservoir; a distribution and regulatory system consisting of mains, branches, distributors, minors and farm-level networks etc. The entire purpose of irrigation management is to achieve performance levels that are in line with the objectives of the irrigation system.

There are different performance objectives of an irrigation system. And, in fact, performance objectives sometimes lack clarity and remain contradictory.

Some important objectives and measures of performance are related to the following:

- increased productivity per unit of water or land;
- improved equity of water distribution among users;
- increased reliability and dependability of supplies to match crop water requirements;
- increased control of water distribution and improved flexibility in operation and maintenance;
- improved environmental stability and maintenance of land and water productivity over time; and
- increased cost-effectiveness of investments in the system.

Management interventions are required to sustain or improve performance levels of any system. A framework detailing performance objectives and elements of managerial interventions that may upgrade performance is given in Table A- 5.1.

Table A-5.1 : Performance objectives of an irrigation system and possible managerial interventions to upgrade performance

	Principal management interventions	
Performance Objectives	*Hardware*	*Software*
I	II	III
A. Productivity		
Improved water and land productivity through reduction of water losses	Canal lining	Improved control and water distribution
Increases in the irrigated area per unit of water	Land leveling, shaping improved irrigation technology	Improved water application
Better crop yields	High yield varieties, improved agricultural pracatices	Agricultural extension
Shifts in cropping pattern	Multiple and relay cropping, shifts to high value crops	Agricultural extension, input and output prices
B. Equity		
Improved equity over head and tail of the main canal system and command	Regulation and control structures	Main system scheduling
Head and tail chaks	On-farm development	Rotational water supply/ warabandi
Large and small farmers Dependability of quantities of water delivered Time delivered	On-farm development Measurement and control structures	RWS / warabandi

I	II	III
C. Flexibility Ability to respond to changing crop and water needs practices	System design, regulatory and control structures	System scheduling and RWS practices
D. Encironmental Stability Short-run and long-run environmental stability through control of areas waterlogged	Main and on-farm drainage	Improved irrigation practices, pricing of water, volumetric charges
Areas rendered saline or alkaline	Drainage, leaching salinity control works, crop selection	

Irrigation by Size of Holding and Source

(Area : '000 ha)

	Marginal	Small	Semi-Medium	Medium	Large	Average/Total	Marginal	Small	Semi-Medium	Medium	Large	Average/Total
	1985-86						1976-77					
I. Net Irrigated Area by Source of Irrigation												
Canal	3095	2865	3514	3775	1858	15107	2234	2268	2909	3197	1515	12123
Tanks	805	638	636	509	196	2784	742	627	730	732	361	3192
Wells	885	1136	1629	2035	1058	6749	806	904	1390	1973	1097	6170
T.Wells	2579	2457	3266	3526	1373	13201	1270	1142	1519	1553	496	5980
Others	698	562	647	515	215	2637	554	484	585	525	224	2372
II. Net Irrigated Area by Source of Irrigation (%)												
Net Irrigated Area	36.63	29.78	26.48	21.98	14.24	24.60	32.55	26.48	22.58	16.63	9.16	18.94
Canal	38.39	37.42	36.25	36.44	39.53	37.39	39.85	41.81	40.78	40.06	41.02	40.63
Wells	10.98	14.84	16.80	19.64	22.51	16.66	14.38	16.66	19.49	24.72	29.70	20.68
T. Wells	31.99	32.09	33.69	34.03	29.21	32.62	22.65	21.05	21.30	16.46	13.43	20.48
Tanks	9.99	8.31	6.58	4.91	4.17	6.88	13.24	11.56	10.23	9.17	9.78	10.70
Others	8.66	7.34	6.67	4.97	4.57	6.52	9.88	8.92	8.20	6.58	6.07	7.95
Total	100	100	100	100	100	100	100	100	100	100	100	100

Source : All India Report on Agricultural Cansus (1976-77) and (1985-86), Dept. of Agriculture and Co-operation, Ministry of Agriculture, Govt. of India.

Area Irrigated by Sources : India (1950-51 to 1982-83)

(Area : Thousand hectares)

Source	1950-51	1960-61	1970-71	1980-81	1985-86*	1986-87*	1987-88*
Govt. Canals	7,158 (34.3)	9,170 (37.2)	11,972 (38.5)	14,450 (37.3)	15,853 (36.6)	15,548 (36.6)	14,990 (34.8)
Private Canals	1.13 (5.5)	1,200 (4.9)	856 (2.8)	842 (2.2)	464 (1.2)	439 (1.1)	484 (1.1)
Tanks	3,613 (17.3)	4,561 (18.5)	4,112 (13.2)	3,182 (8.2)	3,049 (7.2)	2,965 (7.0)	2,806 (6.5)
Tubewells	N.A	135 (0.6)	4,461 (14.3)	9.53 (24..6)	11,866 (22.2)	12,,275 (28.0)	13,156 (30.6)
Others Wells	5,978 (28.7)	7,426 (23.9)	7,426 (23.9)	8,164 (21.1)	8,676 (20.6)	8,552 (20.1)	8,661 (20.1)
Others Sources	2,967 (14.2)	2,266 (7.6)	2,266 (7.6)	2,551 (6.6)	2,609 (6.2)	2,664 (6.3)	2,951 (6.9)
Total (Net irrigated Area)	20,853 (100)	31,103 (100)	31,103 (100)	38,720 (100)	42,079 (100)	42,486 (100)	43,048 (100)

Source : Indian Agriculture in Brief (24th edition), Directorate of Economics and Statistics, Ministry of Agriculture, Government of India.

Area Irrigated by Govt. Canals in West Bengal under Major, Medium and Minor Irrigation Projects

(Area: In hectares)

Project/ Name of the Canals	District	1980-81	1985-86	1990-91	1993-94
I	II	III	IV	V	VI
Total West Bengal		10,17,193	9,97,854	10,68,757	10,70,365
A. Major Irrigation Projects					
I. Mayurakshi Irrigation Project	Birbhum	1,84,911	1,70,763	1,85,876	1,89,381
	Murshidabad	58,374	54,120	28,934	49,596
	Burdwan	15,968	15,890	15,604	15,692
II. Kangsabati Project	Hooghly	16,221	17,891	19,767	19,968
	Bankura	1,26,241	1,36,485	16,577	1,62,769
	Midnapore	1,18,282	1,25,917	1,51,379	1,38,280
III. D.V. & B.I. Project	Burdwan	3,03,337	2,76,940	2,96,759	3,03,302
	Hooghly	83,837	73,214	25,196	73,961
	Howrah	3,957	5,370	4,991	4,821
	Bankura	30,080	31,126	32,014	36,994
B. Medium Irrigation Project					
I. Midnapore Canals	Midnapore	35,110	34,185	22,095	38,091

Contd.

I	II	III	IV	V	VI
II. Suvankar Danra Canals	Bankura	1,760	1,573	–	–
III. Karotha Talms Canals	Jalpaiguri	3,387	1,789	740	2,955
IV. Sanarajara Canals	Purulia	5,532	4,960	4,880	5,520
V. Hinglo Irrigation Project	Birbhum	6,677	6,501	6,941	6,257
VI. Hanumantha Irrigation Scheme	Purulia				
C. Medium DPA Schemes					
I. Bandhu Irrigation Scheme	Purulia	1,902	1,711	1,810	1,880
II. Rupal Irrigation Scheme	Purulia	1,761	1,655	1,509	1,655
III. Dongra Irrigation Scheme	Purulia	751	850	910	956
D. Minor Irrigation Projects	Bankura	3,868	3,510	2,890	3,810
	Midnapore	3,172	2,855	3,640	3,710
	Purulia	3,275	3,378	3,450	3,380
	Jalpaiguri	2,851	2,790	2,540	2,850
	Cooch-Behar	–	2,720	2,600	2,580
	Burdwan	83,510	3,650	84,500	8,75,00
E. Minor DPAP Schemes	Midnapore	1642	1750	18,00	1900
	Bankura	–	1851	1955	2,155
	Burdwan	25100	4,10,50	4,85,50	4,90,50

Source : Directorate of Irrigation and Waterways, West Bengal.

Progress of Irrigation Schemes in West Bengal

(Gross Area Irrigated : '1000 hectares)

Year	*Shallow Tube-Wells*	*DTW*	*RLI*
1980-81	–	2.51	9.11
1987-88	–	1.47	0.40
1988-89	–	2.88	0.68
1989-90		.93	0.72
1990-91	0.45	3.51	0.44
1991-92	2.27	4.53	–
1992-93	4.11	5.13	–

Source : Directorate of Agriculture, Govt. of West Bengal.

Magnitude of Irrigation Subsidies in Agriculture

(Rs. Crores)

Year ending	*Rupees*
1981	651.3
1982	675.6
1983	694.3
1984	729.5
1985	813.4
1986	1030.1
1987	1286.0
Growth rate (Percentage)	10.21

Note : Irrigation subsidies are estimated for recurring costs only.

Source : Gulati, A. (1989) : 'Input Subsidies in Indian Agriculture : A Statewise Analysis : *Economic and Political Weekly*, July 24, 1989.

Gross Irrigated Area as a Proportion of Gross Operated Area Increased from 18.3 per cent in 1960-61 to 33.3 per cent in 1990-91

Area under irrigation	
Year ending	*Gross irrigated area as percentage of gross cropped area*
1961	18.3
1971	23.0
1981	28.6
1986	30.4
1987	31.6
1988	32.9
1989	33.4
1990	33.2
1991	33.3

Source : Agricultural Statistics at a glance, Directorate of Economics and Statistics, Ministry of Agriculture, Government of India.

Area Irrigated by Government Canals in West Bengal

(Thousand hectares)

Year	*Area*
1980-81	1,017.2
1985-86	997.2
1986-87	1,006.4
1987-88	1,039.1
1988-89	1,017.6
1989-90	1,126.2
1990-91	1,096.6
1991-92	1,090,5
1992-93	981.3
1993-94	1,094.8

Source : Report of the Directorate of Irrigation and Waterways, W.B. 1996, p. 136.

Progress of Important Minor Irrigation Schemes in West Bengal

(Number)

Year	*DTW*	*RLI*	*STW*
Upto 1980-81	2332	2697	3769
1985-86	2332	3155	3342
1987-88	2389	3198	3342
1988-89	2528	3239	3342
1990-91	2805	3244	3442
1991-92	3187	3284	3558
1992-93	3831	3274	4479
1993-94	4160	3297	4837

DTW : Deep Tubewell; RLI : River Lift Irrigation; STW : Shallow Tube-wells.
Note : Figures relate to official schemes only of the Dept. of Minor Irrigation, Government of West Bengal.
Source : Directorate of Agricultural Engineering, Govt. of West Bengal

Net Area Irrigated in Undivided India (Average for 1944-45 to 1946-47)

	Canals	*Wells*	*Tanks*	*Others*	*Total*
	I	II	III	IV	V
Area Irrigated (in hectares)	15.2	6.6	3.3	3.1	28.2
Percentage of Total Area Irrigated	53.9	23.4	11.7	11.0	100.0

Source : RIC (Report of Irrigation Commission), Vol. II, p.69.

Irrigated Area as Percentage of Total Area by Size-class of Holdings, All India

Size-class of operational holdings (ha)	*1953-54*	*1961-62*	*1971-72*
(I)	(II)	(III)	(IV)
<1	27.0	34.4	38.4
1-2	25.0	30.5	34.2
2-4	21.0	27.5	30.4
4-6	17.0	23.1	28.5
6-10	13.4	19.5	24.6
10-20	8.7	11.1	11.8
All	14.8	21.0	27.7

Sources :

1. Govenment of India, N.S.S., 8th Round, New Delhi, 1963.
2. Govenment of India, N.S.S., Round No. 17,1968.
3. *Sarvekshana*, Vol. II, No. 2, Oct. 1978, 28th Round

N.B. : We notice a higher percentage of area under irrigation on smaller operational holdings in all these years

Expenditure on Major and Medium Irrigation Projects

Period	*Total expenditure in Current Prices Rs. Million*	*Average Annual expenditure in Constant 1970-71 Prices Rs. Million*	*Average Annual Potential Created Million hectares*	*Expenditure per hectares of Potential Created in Constant 1970-71 Prices Rs.*	*Expenditure as a Percentage of Total Plan Expenditure*	*Expenditure as a Percentage of COP in Market Prices*
(I)	(II)	(III)	(IV)	(V)	(VI)	(VII)
First Plan	3,000	1,366	0.5	2,770	15.3	0.60
Second Plan	3,800	1,500	0.43	3,571	8.1	0.61
Third Plan	5,810	1,504	0.5	3.022	6.8	0.58
Fourth Plan	12,370	2,276	0.5	4,377	7.8	0.54
Fifth Plan	24,420	3,483	1.0	3,483	6.1	0.78
1978-79	9,770	4,934	1.0	4,934	8.8	1.00
Sixth Plan	84,484	7,366	1.1	6,696	8.7	1.13

Source : P. Rogers : *Irrigation and Economic Development : Some Lesson from India*, Cambridge (Mass.), Harvard University, Division of Applied Sciences, 1983.

Development of Irrigation Potential Since 1951[1]

Year	Major/Medium Surface		Minor Surface		Groundwater		Total	
	Million hectares	Annual % Change	Million hectares	Annual % Change	Million hectares	Annual % Change	Million hectares	Annual % Change
(I)	(II)	(III)	(IV)	(V)	(VI)	(VII)	(VIII)	(IX)
1950-51	9.7	–	6.4	–	6.5	–	22.6	–
1955-56	12.2	4.7	6.4	(0.0)	7.6	4.2	26.2	3.0
1960-61	14.3	3.2	6.5	(0.3)	8.3	1.8	29.1	2.1
1965-66	16.6	3.0	6.5	(0.0)	10.5	4.8	33.6	2.9
1968-69	18.1	2.9	6.5	(0.0)	12.5	6.0	37.1	3.4
1973-74	20.7	2.7	7.0	1.5	16.5	5.7	44.2	3.6
1979-80	20.6	4.3	8.0	2.3	22.0	4.9	56.6	4.2
1984-85	32.3	4.0	9.0	2.4	29.0	6.4	70.3	4.5

1.Investment creates irrigation potential when the [illegible] n deliver water to the canal outlet. Irrigation potential is the theoretical gross cropped area that can be irrigated multiplied by the [illegible] intensity. Potential is counted as utilized when farmers actually convey water from the outlet and apply it on their fields.

Source : P. Rogers : *Irrigation and Economic Development : Some Lessons from India*, Cambridge (Mass.) Harvard University, Division of Applied Sciences (1983).

Public Financing for Minor Irrigation Since 1951

Period	*Plan Expenditure*		*Institutional Finance**		*Total*	
	Total Expenditure (Current Prices)	*Average Annual Expenditure (Constant 1970-71Prices)*	*Total Finance (Current Prices)*	*Average Annual Finance (Constant 1970-71Prices)*	*Total (Current Prices)*	*Average Annual (Current Prices)*
(I)	(II)	(III)	(IV)	(V)	(VI)	(VII)
First Plan	660	305	Negligible	Negligible	660	305
Second Plan	1,420	659	190	83	1,610	747
Third Plan	3,280	12,92	1,150	454	4,430	1,746
Fourth Plan	3,260	1,684	2,350	1,126	5,610	2,900
Fifth Plan	5,130	923	6,610	1,185	11,740	2,181
1978-79	6,310	1,177	4,900	1,112	14,110	2,011
Sixth Plan	18,000	1,578	17,000	1,482	35,100	3,060

* Primarily loans to private farmers for tube-wells development.

Source : P. Rogers : *Irrigation and Economic Development : Some Lessons from India*, Cambridge (Mass.) Harvard University, Division of Applied Sciences (1983).

ANNEXURE-6

Impact of Irrigation on Agricultureal Productivity

Project	*State*		*Comparison of areas within and outside the command of selected surface irrigation projects*		
			Percentage Irrigated	*GCA/NSA*	*Output per unit GCA (Rs.)*
Cuttack	Orissa	Command	NA	159.00	3,640.00
		Control	NA	129.00	1,030.00
Mahill	Gujarat	Command	65.00	136.00	1189.00
		Control	34.00	100.00	808.00
Dantimada	Gujarat	Command	63.00	159.00	2172.00
		Control	58.00	126.00	2017.00
Tribeni canal	Bihar	Command	83.00	130.00	563.00
		Control	31.00	111.00	242.00
Cauvery	Tamil Nadu	Command	NA	129.00	328.00
		Control	NA	100.00	81.00

Sources :
1. Satpathy (1984)
2. Adhvaryu (1980)
3. Gadgil (1948)
4. Divakar Jha (1967)
5. Sonachalan (1963)
6. Vaidyaratten (1999)

NA : Not Available

Irrigated Crop Pattern in the Early 1960s, 1970s and 1980s

(Percentage of gross irrigated area)

	1960-61 to 1962-63	*1970-71 to 1972-73*	*1987-88 to 1989-90*
Rice	45.30	37.00	27.70
Wheat	15.30	26.90	31.70
Total Cereals and Millets	72.00	73.60	73.00
Total pulses	7.00	5.00	3.60
Sugarcane	5.60	4.70	5.20
Fruits and vegetables	1.70	2.80	4.10
Cotton	3.60	4.10	4.10
Total oilseeds	1.60	2.90	7.30
Other Crops	8.30	6.70	2.70
Total	100.00	100.00	100.00

Source : GOI, Ministry of Agriculture, Indian Agriculture Statistics (Various issues)

Annexure-6

Estimated Water Utilisation by Use

(Million ha. metres)

	1968-69			1990		
	Gross	*%*	*Net*	*Gross*	*%*	*Net*
Domestic	1.09	2.9	0.47	2.1	2.50	3.5
Industrial	0.26	0.7	0.02	0.1	1.50	2.1
Thermal Power	0.69	1.8	neg.	neg.	1.90	2.6
Agriculture and Live stock	35.69	95.0	21.56	97.6	49.30	91.8
Total	37.55	100.00	22.08	100.00	55.20	100.00

Source : * Chaturvedi, 1976
** Reddy, 1992

Growth of Net Irrigated Area Under Different Sources, All India from 1950/51 – 1993-94

(In milion hectares)

	Canals			Groundwater					Total net irrigated area
	Govt.	Private	Total	Tanks	Tube Wells	Other Wells	Total	Other Sources	
1950-51	7.20	1.10	8.30	3.60	–	6.0	6.0	3.0	20.90
1955-56	8.0	1.40	9.60	4.40	–	6.8	6.8	2.2	22.80
1960-61	9.2	1.20	10.40	4.60	0.20	7.2	7.4	2.4	29.80
1965-66	9.8	1.10	10.90	4.40	–	8.6	8.6	2.5	26.40
1970-71	12.0	0.90	12.90	6.10	4.50	7.4	11.9	2.3	31.20
1975-76	12.9	0.90	13.80	4.0	6.80	7.6	14.4	2.4	34.60
1980-81	14.5	0.80	15.30	3.20	9.50	8.2	17.7	2.6	38.80
1990-91	16.1	0.30	16.10	3.30	14.20	9.90	24.10	2.8	46.30
1993-94	16.6	0.50	17.10	3.20	15.00	10.70	26.50	3.2	50.10

Source : GOI, Ministry of Agriculture, India.

Irrigated Area by Sources as Percentage of Total operated Area by size class

Size Class	1953-54			1981-82			1991-92		
	Surface water	*Ground water*	*Total*	*Surface water*	*Ground water*	*Total*	*Surface water*	*Ground water*	*Total*
	I	*II*	*III*	*I*	*II*	*III*	*I*	*II*	*III*
< 1 ha	18.52	8.04	26.56	19.84	19.05	38.89	19.76	21.84	41.60
1-2 ha	18.02	7.13	25.15	17.64	18.03	35.67	18.07	20.83	38.90
2-4 ha	14.28	6.76	21.04	13.71	18.48	32.19	15.89	20.91	36.80
4-10 ha	9.80	5.36	15.16	11.45	17.70	29.14	13.85	19.02	32.80
> 10 ha	4.39	2.43	6.82	6.30	8.78	15.07	15.55	12.55	28.10
All size class	9.95	4.87	14.82	13.19	16.54	29.73	15.97	19.27	35.40
CV	40.9	32.9	38.2	34.5	23.4	27.3	16.6	17.70	14.40
As per land use statistics	11.97	5.27	17.25	15.21	12.75	27.96	12.93	18.66	31.59

Source : GOI, NSSO. No. 74, *Sarvekshna*, July 1988 and 48th Round, March, 1997.
* GOI, Ministry of Agriculture. Indian Agricultural Statistics

Note : NSS figures related to the ratio of net irrigated area to total operated area, which is larger than net sown area
Groundwater = Tubewells + Wells
Surface water = Canal + Tanks + Others + Not recorded.

Unrecovered costs on Account of Major and Medium Works, Major States

(Rs. m.)

	1987-88[1]	1993-94[2]
Gross revenue	5925	5490
Total cost	52979	129698
Unrecorded costs	47048	124208
Percentage cost (1 as percentage of 2)	12.60	4.4
Public sector plan outlay on irrigation	33469	58540

Source : Srivastava *et al* (1997)
Economic Survey, 1995-96

BIBLIOGRAPHY

Abernethy, Charles L. (1986) : 'Performance Measurement in Canal Water Management' : A Discussion, ODI-IIMI, Irrigation Management Network, August.

Ansari, Nasim (1968) : 'Some Economic Aspects of Irrigation Development in Northern India' in A.M. Khusro (Ed.), *Readings in Agricultural Development*. Allied Publishers, Bombay, pp. 186–239.

——,(1986) : *Economics of Irrigation Rates : A Study of Punjab and Uttar Pradesh*,' Asia Publishing House.

Bardhan, P. (1971) : 'Green Revolution and Agricultural Labourers', *Economic and Political Weekly*, July, 10.

— —,(1999) : 'Rational Fools and Co-operation in a Poor Hydraulic Economy' in *Economic Theory, Essays in honour of Prof. Amartya Sen* edited by Kaushik Basu, pp. 170–180.

Bharadwaj, K. (1974) : *Production Conditions in Indian Agriculture: A Study based on Farm Management Survey*, Cambridge University Press.

——, (1990) : 'Irrigation in India : Alternative Perspectives', Indian Council of Social Science Research, New Delhi.

Binswanger, M.P. (1974) : 'The Measurement of Technical Change Biases with many factors of Production', *The American Economic Review*, Vol. LXV, No. 3, Dec. 10, pp. 964–76.

Biswas, A.K. *et al.* (1993) : *Water for Sustainable Development in the 21st Century*, Oxford University Press, New Delhi.

Bottral, Anthony F. (1977) : 'Skills, Functions, Organisational Forms in relation to Project and System Characteristics'. Discussion paper 1/77/33. ODI, London.

——, (1981) : 'Comparative Study of the Management and Organisation of Irrigation Project', World Bank Staff Working Paper, World Bank No. 458, May, pp. 15–18.

Bromley, Daniel W. (1982) : 'Improving Irrigated Agriculture : Institutional Reform and the Small Farmer', World Bank Staff Working Paper' No. 531, The World Bank, Washington D.C.

Byres, T. (1971) : 'Dialectics of Green Revolution', *South Asian Review*, Vol. 5, No. 2, pp. 99–116.

Carruthers, Ian (1987) : 'Irrigation Under Threat - A warning Brief for Irrigation Enthusiasts in Irrigation Design for Management,' *Asian Regional Symposium*, Vol. 11, Hydraulics Research, Walling Ford, England, p. 39.

Casser, Robert and Vijay Joshi (1996): *India : The Future of Economic Reforms*, Oxford University Press, New Delhi.

Chambers, Robert (1988) : *Farmers above the outlet : Managing Canal Irrigation*, Oxford and IBH Publishing Co., New Delhi.

——, (1988) : *Managing Canal Irrigation*, Oxford and IBH Publishing Co. Pvt. Ltd., New Delhi.

— —, (1990) : 'Men and Water : The Organisation and operation of Irrigation' in B. Farmers (ed.) *Green Revolution and Technological Change in Rice-Growing Areas of Tamil Nadu and Sri Lanka.*

Chaturvedi, M.C. (1976) : *Second India Series : Water*, MacMillan, New Delhi.

Chopra, K., G. Kadekor and M.N. Murthy (1988) : 'People's Participation and Common Property Resources' *Economic and Political Weekly*, No. 24.

Chow, C. (1960) : 'Tests of Equality Sets of Co-efficients in two linear regression' in *Econometrica*, Vol. 28, No. 3, July, pp. 591–605.

Colenam, J. (1990) : *Foundation of Social Theory*, Harvard University Press, Chambridge, Mass.

Desarade, H.M. (1995) : 'Crisis of Abundance and Paucity' in *The Economic Times*, Calcutta, October 23.

Dhawan, B.D. (1983) : 'Role of Irrigation in Stabilising Income and Farm Output in India', Working Paper, Institute of Economic Growth, Delhi.

——, (1983) : 'Equity in Sharing Gains from Irrigation', Working Paper, Institute of Economic Growth, Delhi.

——, (1983) : 'Impact of Irrigation on Instability in Farm Output : A Case Study of Tamil Nadu State', in *Economic and Political Weekly*, 18(31), pp. 109–16.

——, (1985) : 'Irrigation impact on Farm Economy', in *Economic and Political Weekly* 20 (39), Sept. 28.

——, (1986) : 'Irrigation and Water Management in India' : Perceptions of Problems and Their Resolutions', in *Indian Journal of Agricultural Economics*, Vol. 41. No. 3, July Sept., pp. 271- 81.

Easter, William K. (1977) : *Water Resources Problems in Developing Countries*, Economic Development Centre, University of Minnesota, p.80.

Eastser, K.W. and Palanisami, K (1986) : 'Tank Irrigation in India and Thailand: An Example of Common Propery Resource Management, University of Minnesota', Department of Agricultural and Applied Economics, Staff paper, August.

Ekstein, Otto (1985) : *Water Resources Development - The Economics of Project Evaluation*, Harvard University Press, Massachusetts, p. 52.

Frankel, F.R. (1971) : *India's Green Revolution – Economic Gains and Political Costs*, Oxford University Press.

Gadgil, D.R. (1948) : 'Economic Effects of irrigation', Report of Survey of the Direct and Indirect Benefits of Godavari and Pravara Canal, Pub. No. 17, Gokhale Institute of Economics and Politics, p. 22.

Ghai, Dharam (1988) : 'Participatory Development – Some Perspectives from Grassroot Experience', UNRISD, Geneva.

Ghosh, M.G. (1996): Servey Report on Mayurakshi canal, *Irrigation Project*, Santiniketan, West Bengal.

Hanumantha Rao, C.H. (1976) : Growth of Irrigation in India: An Outline of Performance and Prospects' in *Seminar Series XIII*, Bombay, Indian Society of Agricultural Economics.

Hanumantha Rao, C.H. , Ray, S.K.and Sibbarao, K. (1988): *Unstable Agricultlure and Droughts*, Vikas Publishing House Pvt. Ltd.

Hardin (1978) : 'The Tragedy of the Commons', *Science*, Vol. 162. Washington D.C.

Hart, H.C. (1978) : 'Anarchy, Paternalism or Collective Responsibility under the Canals', *Economic and Political Weekly*, 13 (15), pp. 125-34.

Hashim Ali, Syed (1980) : 'Integrated Water Management Above and Below the Outlet', Paper Presented at the Workshop on Warabandi at the Administrative Staff College of India, Hyderabad.

Hicks, J.R. (1964) : *Theory of Wages*, Macmilan Co., Ltd., New Delhi, Second Edition, pp. 121-23.

Hirasima, S. and Gooneratna, W. (1990) : *Irrigation and Water Management in Asia*, Sterling Publishing Private Ltd.

Hunt, E. and Hunt, R. (1976) : 'Canal Irrigation and Local Social Organization', *Current Anthropology*, Vol. 17.

Indian Society of Agricultural Economics, (1975) : Proceedings of The Seminar on the Role of Irrigation in Development of India's Agriculture, Bombay.

Ishikawa, S. (1974) : *Economic Development in Asian Perspectives*, Institute of Economic Research, Hitotsubhashi University, Tokyo.

Jairath, J. (1985) : 'Technical and Institutional Factors in Utilisation of Irrigation - A case Study of Public Canals in Punjab', *Economic and Political Weekly*, 20 (13), pp. A2-10.

Kanwar, J.S. , (ed) (1986) : 'Water Management : The Key to Developing Agriculture, Indian National Science Academy, New Delhi.

Kumar, Vinita (1997) : *Economic Growth and Rural Poverty : The Indian Experience 1960-95*, Concept Publishing Company, New Delhi.

Ladejinsky, Wolf (1969) : 'Green Revolution in the Punjab', *Economic and Political Weekly*, December.

Lenton, Roberto (1988) : Strategy Development in IIMI Irrigation Management Network, Paper 81/1e Overseas Development Institute, London, p.11.

Majumdar, D. (1975) : *District Gazetteers,* West Bengal, Birbhum District.

Meinzer - Dick, Ruth Svendsen, Mark (ed) (1991) : Future Directions for Indian Irrigation : Research and Policy Issues, International Food Policy Research Institute (IFPRI), Washington D.C.

Misra, Sibranjan (1989) : 'A Study of Impact of Mayurakshi Canal Irrigation Project on Cropping Pattern, Employment and Income Distribution', Report of the Project sponsored by Indian Council of Social Science Research, New Delhi.

Moorti, T.V. (1976) : 'Impact of Different Sources of Irrigation on Input-Output Relations, Crop Pattern and Farm Policies' in *Seminar Series*, XIII, Indian Society of Agricultural Economics, Bombay.

Miller, D. (1979) : *Self-help and Popular Participation in Rural Warer System*, Development Centre, OECD, New Delhi.

Nadkarni, M.V, *et al.* (1979) : *Impact of Irrigation - Canal, Well and Tank Irrigation in Karnataka*, Himalaya Publishing House, Bombay.

Narain, D and Roy, S. (1980) : 'Impact of Irrigation and Labour Availability on Multiple Cropping :A Case Study of India, Research paper No. 20, International Food Policy Research Institute, Washington D.C.

National Council of Applied Economic Research (1959) : 'Criteria for Fixation of Water Rates and Selection of Irrigation Projects', Asia Publications House, New Delhi.

Navalawala, B.N. (1997) : 'Modernization of Irrigation Schemes in India' : FAO Publications, Water Report No. 22.

Oakerson, R.J. (1986) : 'A Model for the analysis of Common Property Problems', in National Research Council, Proceedings of the Conference on Common Property Resources Management, Washington D.C., National Academy Press.

Open Von, C.M. *et al.* (1983) : Alternatives for Improving Small Scale Irrigation System in Watershed in India, ICRISAT, Hyderabad.

Ostrom, E. (1990) : *Governing the Commons : The Evolution of Institutions for Collective Action*, Cambridge University Press, New York.

Pant, Niranjan (1982) : 'Major and Medium Irrigation Projects : Analysis of Cost Escalation and Delay in Completion' in *Economic and Political Weekly*, 17(26), A 34–43.

— —, (1984) : *Productivity and Equity in Irrigation System*, Ashish Publishing House, New Delhi.

— —, (1986) : 'Farmers' Organisation in Large Irrigation Projects', in *Economic and Political Weekly*, December 27, vol., XXI, No. 5d2, pp. A175–78.

Parthasarathy, G. (1974) : *Agricultural Development and Small Farmers*, Vikas Publications, pp. 26–36.

Patel, A.S. (1981) : 'Irrigation : Its Employment Impact in the Command Areas of Medium Irrigation Projects in Gujarat', : *Indian Journal of Agricultural Economics*, 36(4), pp. 20–30.

Prasad, K.N. (1972) : *Economic Benefits in the Kosi Command Area*, A.N. Sinha Institute of Social Studies, Patna.

Raj, K.N. (1960) : *Some Economic Aspects of the Bhakra Nangal Project*, Asia Publishing House, Bombay.

Rao, K.H. and Sastry, K.R. (1997) : 'Decentralized Planning Framework for Public Interventions'. *Asia Pacific Journal of Rural Development*, Vol. VII, Dec., No. 2, pp. 33–42.

Reidinger, K.B. (1974) : 'Institutional Rationing of Canal Water in Northern India : Conflict Between Traditional Pattern and Modern Needs', *Economic Development and Cultural Change*, Vol. 23, No. 1, October, pp. 103–04.

Repetto, R. (1986) : Skimming the Water : Rent Seeking and the Performance of Public Irrigation Systems, Research Report 4, World Resources Institute, Washington.

Report of National Commission of Agriculture (1976) : Part II, Ministry of Agriculture and Irrigation, Government of India, New Delhi.

Report of Irrigation Commission, Vol. I, II and III, (1972), Ministry of Irrigation and Power, Government of India, New Delhi.

Rogers, P. (1983) : *Irrigation and Economic Development : Some Lessons from India*, Cambridge (Mass.) : Harvard University, Division of Applied Sciences.

Rogers, P. and Smith, D.V. (1970) : 'The Integrated Use of Ground and Surface Water in Irrigation Projects', *American Journal of Agricultural Economics*, 19/3.

Runge, C. (1986) : 'Common Property and Collective Action in Economic Development', *World Devlopment* 14.

Sampath, Rajat K. (1992) : 'A farm Size-wise Analysis of Irrigation Distribution in India', *The Journal of Development Studies*, Vol. 29, No. 1, October.

——, (1984) : 'Income Distribution Impacts of Irrigation Water Distribution Policy', *Water Resources Research*, Vol. 20, No. 6, pp. 647–654.

Sastry, K.R. (1997) : ' Factors Affecting People's Participation', *Asia-Pacific Journal of Rural Development*, Vol. VII, No 1, July, pp. 30–35.

Sargent, M. (1982) : *Agricultural Co-operation*, Aldershot, Gower.

Satpathy, T. (1984) : *Irrigation and Economic Development*, Ashish Publishing House, New Delhi.

Sau, Ranjit (1971) : 'Resource Allocation in Indian Agriculture', *Economic and Political Weekly*, September, p. 107.

Seckler, David (1981) : *The New Era of Irrigation Management in India*, The Ford Foundation, New Delhi, India.

Seabright, Paul (1997) : 'Is Co-operative Habit Forming' in The *Environment and Emerging Development Issues* (Vol. 2) by Partha Dasgupta et al. (ed.), Clarendon Press, Oxford.

Selvarajan, S. and Subramaniam, S. R. (1984) : 'Economic Impact of Resource Use Optimization and Water Augmentation in Farms of Parambikulam Aliyar Project Region', *Indian Journal of Agricultural Economics*, 36(1), pp. 89–100.

Sen, Amartya (1987) : *On Ethics and Economics*, Basil Blackwell, Oxford.

Sengupta, N. (1991) : '*Managing Common Property : Irrigation in India and Philippines*', Sage Publications, New Delhi.

— —, (1982) : *Tank Irrigation in Gangetic Bihar*, A.N.S. Institute of Social Studies, Patna.

Shukla, Lakshmi and Ram Kumar Gurjar (1991) : *Canal Irrigation Management Problems - Time and Use Relationship*, Agricole Publishing Academy, New Delhi.

Sharma, S. (1984) : High Costs of Irrigation, *Indian Express*, New Delhi - November 4.

Singh, K.K. (1991) : 'Farmers' Participation in Irrigation Management : Basic Issues' in *Farmers in the Management of Irrigation systems* by K.K. Singh (ed.), Sterling Publishers Pvt. Ltd. New Delhi.

Silgleton, S. and Taylor, M. (1992) : 'Common Property, Collective Action and Community', *Journal of International Politics*, 43, pp. 309–24.

Sirohi A.S. (1985) : 'Impact of Agricultural Subsidies and Procurement Prices on Production and Income Distribution in India', Presidential Address at the Annual Conference of the Indian Society of Agricultural Economics, 1985.

Singh, Kartar (1991) : Managing Common Pool Irrigation Tanks : A Case Study in Andhra Pradesh and West Bengal, Institute of Rural Management, Anand, India.

Singh, J.P. *et al.* (1979) : 'New Seeds, Adoption and Yield', Institute of Social and Economic Change, p. 17.

Stone, Ian, (1979) : 'Canal irrigation and Agrarian Change : The Experience of the Ganges Canal Tract, Muzaffarnagar District, 1980–1900' in K.N. Chowdhury and C.J. Dewery (ed.), *Economy and Society*, Oxford University Press, pp. 86–111.

— —,(1984) : *Canal irrigation in British India : Perspectives of Technological Changes in a Peasant Economy*, London, Cambridge University Press.

Suresh, K.A. and Joseph, M. (1990) : *Public Participation in Rural Development : A Case Study of Non-Governmental Organization in Kerala*, College of Co-operation, Kerala Agricultural University, Thrissar.

Tang, S.Y. (1991) : 'Institutional Arrangement and the Management of the Common Pool Resources', *Public Administration Review*, 51/1, pp. 42–51.

Taylor, M. (1987) : *The Possibility of Cooperation*, Cambridge University Press,

Thavaraj, S.H.C. (1982) : 'The Importance of Integrating Non-engineering Aspects in Irrigation System Design' in Donald C. Taylor and Thomas H. Wickham (ed.) *Irrigation Policy and the Management of Irrigation System in South-East Asia*, The Agriculture Development Council, Inc., Bangkok, 15.

Thorner, Daniel (1959) : 'The Weak and the Strong on the Sarda Canal', in *Land and Labour in India*, Asia Publishing House.

Uphoff, N. (1986) : *Improving International Irrigation Management with Farmer's Participation*, West View Press, London.

——, (1992) : 'Local Institutions for Sustainable Development', Gatekeeper Series, No. 31, International Institute for Environment and Development, London.

Vaidyanathan (1983) : 'Water Control Institution and Agriculture - Comparative Perspective', Working paper No. 178, Centre for Development Studies, Trivandrum, October.

——, (1982) : 'Equitable Distribution of Irrigation Water', *Kurukshetra*, Vol. XXX, No. 9.

——, (1999) : *Water Resource Management : Institutions and Irrigation Development in India*: Oxford University Press, New Delhi.

Venkata Reddy, N. (1990) ; *Dynamics of irrigation Water Management*, Ashish Publishing House, New Delhi.

Vohra, B.B. (1975) : 'Land and Water Management Problems in India', Training Division, Department of Personnel and Administrative Reforms, Cabinet Secretariat, New Delhi.

——, (1988) : 'Irrigation Water Management : Need for a Fresh Look', *Proceedings of the International Conference on irrigation System, Evaluation and Water Management,* Wuhan University of Hydraulic and Electric Engineering, Wuhan, Hubei, People's

Republic of China.

Wade, R. (1976) : 'Performance of irrigation Projects', *Economic and Political Weekly*, 11(3), pp. 61–66.

— —, (1979) : The Social Response to Irrigation : An Indian Case Study', *Journal of Development Studies*, 16(1), pp. 3–26.

— —, (1982) : 'Group Action for irrigation', *Economic and Political Weekly*, 17(39), pp. A 103 -06.

Wade, R. and Chambers, R. (1980) : 'Managing the Main System : Canal Irrigation Blind Spots', *Economic and Political Weekly*, Vol. XV, No. 39.

Whitcombe, E. (1972) : *Agrarian Conditions in Northern India*, Vol. I : The United Provinces under British Rule 1860–1900, Thomson Press (India), New Delhi, 1971.

Wills, I.R. (1971) : 'Green Revolution and Employment', *Economic and Political Weekly*, March 10.

Wittfogel, K.A. (1957) : *Oriental Despotism : A Study of Total Power*, New Haven : Yale University Press.

Wolfer, W. Hans and Charles, M. Burt (1997) : 'Concepts of Modernization : Modernization of Irrigation Schemes : Past Experience and Future Options', FAO Publications, 1997/22, Water Report 22, Bangkok.

INDEX